# Developing Mathematics with UNIFIX®

by Dr. Paul Swan and Geoff White

Order Number 2-177
ISBN 1-58324-244-9

A B C D E F 10 09 08 07 06

395 Main Street
Rowley, MA 01969
www.didax.com

# CONTENTS

# Foreword

For many decades, educators have recommended the use of manipulative materials to assist young children in their learning of mathematics. The advocacy of educators such as Maria Montessori, Zoltan Dienes and Catherine Stern encourages a wide acceptance of the use of manipulative materials, especially in elementary school classrooms. Once, it was felt that simply giving students manipulatives to use in mathematics lessons would be enough to develop an understanding of mathematical concepts. This is not true. Manipulatives in and of themselves do not teach—skilled teachers do.

This series—Developing Mathematics—is designed to help teachers who are trying to make the most of students' experiences with manipulatives. It is better to use a few well-chosen manipulative materials rather than an array of bits and pieces so students will have an adequate supply of pieces. We recommend a lot of a little rather than a little of a lot when it comes to working with manipulatives. Nothing is more frustrating than not having enough to finish creating a design or building that masterpiece. It is also important that sufficient materials are available to allow models to be left on display in the classroom.

## Why use manipulatives?

When used as part of a well thought-out lesson, manipulatives can help students understand difficult concepts. The key to good use of manipulatives is for teachers to have a clear goal in mind. This will help maintain the intention of the lesson and focus responses to any questions asked during the lesson. Teachers will have a clear idea of what to look for when observing students using manipulatives.

We have observed how students experiment with ideas willingly. If satisfaction with an idea is not achieved, students will seek another solution. We do not see this happening as frequently when students are expected to work with abstract statements such as equations and written problems.

The skilled use of manipulatives will enhance mathematics outcomes. Poor use many be detrimental to student attainment. This series of books is designed to ensure skilled use of manipulatives in the classroom.

## Is there a difference between a mathematics manipulative and a mathematics teaching aid?

We believe there is a big difference between the two types of materials.

A child can interact with and even take control of a good mathematical manipulative; whereas a teaching aid tends to control the learning experience. Too often, a teaching aid is used as a telling support rather than a learning support and experience has taught us that "telling" is not a very successful method of teaching mathematical ideas.

## How will I know whether the students are learning anything?

Observe the students as they work with the manipulatives. Don't worry if they solve a problem in a way different from what you expected. Ask questions. Encourage students to explain their thoughts or write about their experience.

## What evidence can I show that students are learning?

Some teachers are concerned about the lack of written evidence to substantiate learning when manipulatives form a large part of the lesson. There are several ways a student might record his/her findings:

- writing about the experience,
- sketching or drawing any models produced,
- photographing any models produced,
- presenting to students in other classrooms and
- maintaining a learning journey log book.

Actually, when preparing this type of learning evidence, students have a wonderful opportunity to reinforce their own learning.

## How do I manage the use of manipulatives?

Some teachers worry that students will only play with the manipulatives and not pay attention, or worse still, begin to throw the material around. These are genuine fears which will decrease as experience, both by the students and teacher, increases.

The first time you introduce a manipulative, allow time for the students to explore it. Set some simple rules and limits for the way the material is used and enforce these early on. Students will soon learn to respect the material. Throughout this book, we present management ideas. We encourage you to adopt them as your own.

# More than Fifty Years of Unifix®

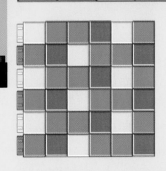

In classrooms throughout the world, Unifix Cubes are used as a key manipulative material.

The cubes were first manufactured in 1953. The designer, Charles Tacey, was a partner in Philip and Tacey, a company which has been in existence since 1829. At that time, beads were made of wood and to hold them together they had holes in them through which a lace could be threaded. Even though he left the threading hole, Tacey created the interlocking idea, an idea which has been copied and adapted by designers of a wide variety of interlocking blocks ever since.

Made of polyethylene plastic, the blocks are sturdy and make very little noise on laminated table tops. The softness of the cube is user-friendly and young children can connect the cubes easily.

The pioneering *Mathematics Their Way* (1975) by the late Mary Baratta-Lorton, made extensive use of Unifix Cubes. In 1999, Mary's colleague, Kathy Richardson, created a series of four books under the general heading of *Developing Number Concepts* (1999) which feature the extensive use of Unifix Cubes in a systematic and developmental manner. Many other books which provide activities using Unifix Cubes have also been produced.

These blocks are called Unifix because they join or fix at one point. The cubes form a connected series easily. Creative pattern-making is fostered by using Unifix Cubes.

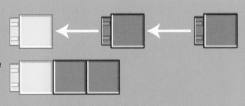

A wide range of materials has been produced to support the use of the cubes. *Developing Mathematics with Unifix* provides the classroom teacher with a comprehensive coverage of the many applications of the familiar range of Unifix materials.

By naturally integrating these materials into the students' learning processes, successful mathematical learning will be achieved. Unifix materials foster developmental experiences in the mathematical learning process when the students learn by doing in an environment that encourages them to think, reflect and discuss the concepts being developed.

# A Guide to Using Unifix® Materials in the Classroom

## Sufficient quantities

We recommend that a classroom has access to at least two containers of 1000 cubes at any one time. Students relax when they know there is sufficient material available. This also permits models to be left on display for extended periods. Some teachers place a collection of cubes in drawstring bags or plastic containers to speed up distribution in the classroom.

Limit the length of time the students use the cubes.

The idea of 2000 cubes is not really extravagant because the cubes need to be in the classroom for about three weeks at a time. In a 10-week term, that means three classrooms can have exclusive use of the material for three straight weeks. After three weeks, another set of material, pattern blocks, for example, can be rotated into the classroom for a three-week period. This system works very well. Our motto: *Three weeks on: six weeks off*.

During the six-week break, the student's brain carries out a great deal of assimilation. When the cubes return, the students are ready to go again. The same comment can be made about other materials.

## Easy access at all times

Store the cubes in plastic tubs that can be carried easily. Keep them in a dedicated space in the classroom so students can have access to them at any time. Models of solutions and creations can be displayed in easily accessible places. While school is in session, the cubes will be in classrooms, so no special storage facilities are needed. This means the manipulatives are out in classrooms, where they belong, rather than collecting dust in the storeroom.

## Take-home bags

Some schools encourage students to take home a draw-string bag containing 50 or so cubes instead of a reading book. A small card may explain an activity which the student teaches the parents. So, rather than a parent demanding, "What did you do in mathematics today?", a student can show the cubes and say "Look, I can do a snap-clap pattern." Do not fear you will lose the cubes. The students will have developed a pride in them and will take very good care of the cubes.

## Keep the containers and cubes clean

Washing in warm soapy water once or twice a year will keep the cubes looking new. Rinse the container in disinfectant and leave a light cloth with a sprinkling of disinfectant in the bottom of the tub.

The mathematical world of a child does not develop in a straight line or a predictable, sequential manner. In fact, it can be said the whole procedure is pretty messy. Ideas shoot into the brain from all directions in no coherent order. It is the function of the brain to find an order which suits the child at that time.

Here seems a paradox. As educators, we know that for most learning to take place, certain stages of development need to have been achieved by a human being. It seems that development is independent of our learning ideas. Some would say there is no point in attempting a certain piece of learning unless the learner has reached a certain stage of development. Others would suggest that providing learning experiences before the child is ready could cause long-term damage to the child's capability and willingness to learn.

# Challenging
## the Step-by-Step Approach

As a teacher there will be times when you will find the ideas you teach are only half understood and you become frustrated and perhaps even say things like, "That child is just not learning." Maybe that child is not ready to learn what you want him/her to learn! Suddenly, after sharing some other activities and, often, on some other day, that child will demonstrate the "aha!" factor—"I understand." Why should a young child's experience be any different from yours? You get "ahas" at any time and in any place.

When preparing a text such as this, ideas can be presented in a carefully laid out plan. But this does not mean that the student learns these things in that order and it definitely does not mean that you have to present the ideas in the order offered here.

All these ideas have been well tried with children and their introduction to the young learners can be just another part of their real life experiences. Give the students a chance to create a whole picture of mathematics, rather than force isolated sections on them.

# Introducing
# Learning Stations

We encourage the establishment of Learning Stations in a classroom to promote the development and strengthening of ideas. In groups of three to six, class members may congregate around a challenge and spend **their** time on the idea.

Imagine! At any one time in your classroom, several widely different activities will be involving the students. We have seen this type of classroom organization used in many elementary schools. An interesting benefit of this classroom management technique is that the students take ownership of the Learning Stations. And, from your point of view, you will be very satisfied with the amount of learning taking place.

*Students need time to experiment with ideas.*

You can integrate activities from other curriculum areas. A colleague organized all lessons using Learning Stations as the foundation of classroom management. Each morning, he greeted the members of the class and frequently presented a lesson to the whole group. Often, this lesson served to initiate a new Learning Station. At any time, there were about eight Learning Stations in the room, three or four of them being introduced during the week. Sometimes, Learning Stations were removed and then re-installed weeks later to help reinforce learning experiences.

# Key Elements in a Learning Station

- Students keep a record of their activity in a simple Learning Station logbook. We do not insist on copious records: reminders are written to encourage conversation. The maintenance of the classroom Learning Stations is the responsibility of the class members. Some students may find it difficult to interpret the instruction card but will find classmates willing to help out.

- An activity space or a table or pair of desks may be dedicated to a Learning Station. We often cover the activity space with a bright table cloth. It looks inviting and the students like it.

- Appropriate manipulative materials are kept at the Learning Station. These will include the correct manipulative, recording materials (paper and pencils) and simple instructions.

- Throughout this book, pages have been dedicated to Learning Station ideas. Simply photocopy then laminate the page and display it on a card holder. Students learn very quickly to obey the number rule, which indicates the maximum number of students to be at a Learning Station at any one time.

- The teacher's role is transformed from one of up-front director to one of a sharing participant.

# Towards Mathematical Abstraction

Zoltan P. Dienes, the mathematics educator/psychologist, saw that children acquired understanding gradually and only after sufficient directed play had been experienced. Today, this approach may be known as constructivism; that is that a learner shapes his/her learning via interaction with the environment. The teacher has a vital role in shaping that learning environment.

No matter what activity you present to the students, they will experience the stages of *Discover* and *Talk*. Students need time to find out what it is all about. For some, it may be a new experience altogether. Consequently, more time will be spent at the *Discover* and *Talk* stages of development. In contrast, some students may be well on the way to formalizing a concept; hence they will be engaged at the *Explain* and *Symbol* stages.

We have observed students passing through stages of mathematical understanding and capability. We are using an acronym for this—DTES—and have applied it throughout this collection of ideas on how to best use Unifix® materials. Where possible, for your guidance, we have indicated an approximate developmental stage for the activities.

## Discover ➡️ Talk

A child experiences his/her environment: the child sees, hears, feels, tastes, smells and handles. In fact, all learning begins at this point. The wider the experience, the richer will be the language development.

A teacher does not need to direct this experiential stage—there is no harm in suggesting ideas and, if the students run with them, good! On the other hand, this is a time when the teacher will be able to observe the students, note their developmental stages and talk to them about their ideas.

Do not rush this stage!

A child develops the spoken language to describe and communicate that experience. The interaction with peers and significant others, such as parents and teachers, strengthens that development.

## Symbol ⬅️ Explain

Within the written symbol, there is a huge amount of knowledge (discover, talk and explain). To make this symbol and to comprehend other symbols is a very sophisticated achievement and must never be forced or "fast-forwarded."

There are many ways in which children will explain their ideas—perhaps in speech, pictures, writing, or actions. But, whatever the type of representation, the child recognizes the association with the original ideas.

## Using DTES as a guide

On various activities we will use the DTES symbol to provide a rough guide of the developmental levels involved in the activity.

**D T E S**

Students will lead the activity as they discover various ideas and suggest conclusions.

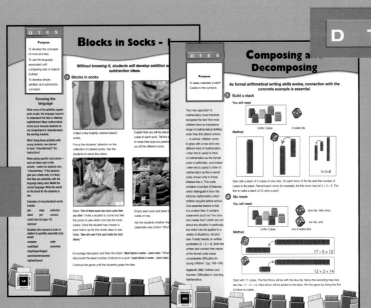

**D T E S**

Students have developed many conceptual ideas. Now, they are ready to present them formally.

**Purpose**

To help familiarize students with Unifix Cubes and the language associated with their use. As a result, the students will learn the names of the colors of the cubes.

# Becoming Familiar with the Cubes

*Even though instinctively you may feel otherwise, students need play and more play so they are absolutely at ease with the Unifix® Cubes. Students' discovery experiences take many forms.*

## The Unifix colors are:

white

light blue

yellow

red

green

orange

brown (tan)

black

(dark) blue

maroon*

\* Note: The color maroon may be confusing to some students. It can be called several names. Some teachers have suggested removing the maroon cubes in the early stages of becoming familiar with the cubes.

Also, it has been noted that some students have displayed conflict with mathematical ideas because they are struggling with the names of the colors.

Some teachers introduce the cubes to the students after removing the last five colors in the above list. They claim it is initially less confusing for the young students. Later on, these cubes will be introduced into the collection. You decide what is best for your students.

**1** ## Loosely directed play

Ask the students to:

• make stacks of different colors.

• find all the cubes of their favorite color.

Students will combine forces to stack the cubes into long snakes or trains.

**2** ## More controlled play

Ask the students to:

• sort the cubes into four stacks.
  *Why did you make your stacks like that?*
  (You will learn a lot as you listen to the students' replies.)

• make a tall stack. *Is your stack the tallest?*

• grab as many cubes as you can in your left (right) hand. *Who has the most cubes?*

• find the container which holds the least (or the most) number of cubes from a variety of plastic containers.

• stand above paper circles and squares that have been placed on the floor and drop a handful of cubes from a height. Then ask them to count the cubes that fall on the shapes.

**3** ## Cubes in cups

Set up paper cups with numbers printed on them. For example:

Place that number of cubes in the cup.

Order from: least to most

most to least

odd/even numbers

**4** ## Stacks

Ask students to:

• make two-block stacks with two colors.

• make three-block stacks with three colors.

• make different three-block stacks using the same colors.

**5** ## More or less – Same or different

One stack of Unifix Cubes (no more than five) is placed in front of a group of students. Taking turns, the students throw a soft 6-sided dot die. The number of dots will determine the number of cubes that will be in the student's stack. Score a point if he/she has more than the sample stack. If the number of dots is the same as the model, the student can build a model the same as the sample stack, and score a bonus of three points.

Each student takes a turn to say *"My stack is more (or less) than the model."* No points for less!

 After a few throws, change the scoring rule to earn a point for less. Later, you can introduce, *"My stack is two less; I score two points."*

The concept of "different" may be gently introduced by encouraging the students to say, "My stack is bigger than the model by two cubes," or "I have two less than the model." Reinforce the ideas: *"I have a stack. It is three more (or less) than yours."* Don't let the student see your stack. *"Make my stack."*

## Key concepts

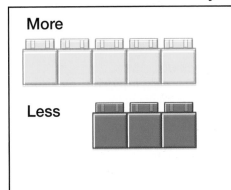

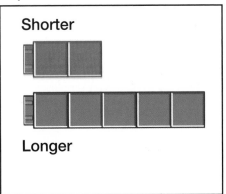

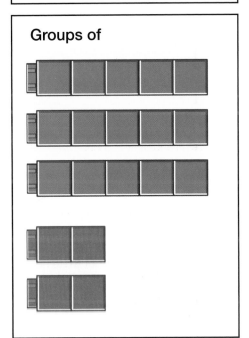

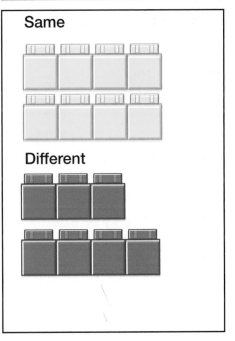

These ideas take time and experience to develop in a young student. Relate these words and concepts to other areas of activity; for example, compare patterns of clothing, size of drink containers and number of students present in class.

While the students "play" they will discover many key mathematical ideas such as classifying. Sometimes the students will lack the language to describe what they notice or discover. Encourage students to talk to one another about what they discover. Often this discussion will be informal and as such, the language may not be sophisticated. As the teacher, feel free to join in with these group discussions as you move around the room. Be careful, however, not to dominate the discussion or impose on the students

Where possible, try to restate the ideas expressed by the students. In a whole-class discussion at the end of the lesson, encourage students to share their findings. You may decide to introduce some more formal mathematical vocabulary. For example; the concepts of *same* and *different* are quite complex. Two stacks of cubes may be the same and yet different. One stack may be made of 5 red cubes, while another may be made of 5 green cubes. Each has the *same* number of cubes and is the *same* length, but its colors are *different*. Frequently, confusion arises when students use different attributes to determine sameness and difference.

# Listening Skills are Important

## Purpose

To learn how carefully students are listening and thinking about what they are hearing. Be aware that some of these challenges may be open to different and yet correct interpretations. Discuss ideas in depth.

## Listening and hearing are different

Many students do not hear what we want them to hear. Also, many students hear what we say and then reinterpret it to suit their own experience.

As teachers, we must resist blaming students for these misunderstandings. Take into account the learning experience, the maturity, the health and the developmental stage of the student.

To improve the mathematics results in our classrooms, we need to give the students lots of practice at listening to instructions. Students need to be given time to develop their listening skills and real time to think things out. That may mean several days may pass before a response or solution is forthcoming. That's OK! There is no need to rush. Too much mathematics failure can be attributed to poor listening/ hearing skills.

### Do your students really hear what you say when you set a challenge or discuss ideas?

Listen very carefully to students' responses. Work carefully to discover what they mean, not what you want them to mean. Learn to respond to the student supportively with replies such as:

- *Did I hear you say …?*
- *I did not get that. Try again to tell me …*
- *I like what you said, but what if you looked at it in this manner …?*
- *Why did you say that?*

Do not accept "Yeah!" or "I dunno!" as a response. Encourage the students to provide thoughtful answers, even though there may be a bit of a struggle to get there!

Avoid pandering to the student. It is essential that we take students to their intellectual edge where new thinking is part of their everyday life. The developing brain thrives on a challenge to its present state of mind.

### Guide to using this photocopiable resource

Use the challenges opposite to create Learning Stations where students respond to the challenge on the card.

## Listening skills cards

### Card 1
Students have learned the definition of a stack, so a huge variety of stacks will be offered.

### Card 2
This statement allows for more than two colors to be used.

### Card 3
In this example, the rules are strict.

### Card 4
Another interpretation of multi-colored stacks: no color may be repeated in any one stack.

### Card 5
While this appears to be self-explanatory, some students will show difficulty in holding the rules together as one idea.

### Card 6
Again, while seeming simple, the whole idea needs to be worked out in the mind.

Other variations:

- Where possible, colors must have an equal number of cubes.
- Use three colors.
- Use two colors but white has the most cubes.

### Card 7
Clearly, three or more colors would be permissible and the use of only two colors is acceptable.

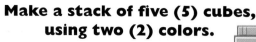

## Make a stack of five (5) cubes, using two (2) colors.

Listening Skills Learning Station Card #2

## Make a stack of five (5) cubes. Each cube must be a different color.

Listening Skills Learning Station Card #4

# Make a stack of five (5) cubes.

Didax Inc.®

Listening Skills Learning Station Card #1

# Make a stack of five (5) cubes, using two (2) colors.

Didax Inc.®

Listening Skills Learning Station Card #2

# Make a stack of five (5) cubes, using only two (2) colors.

Didax Inc.®

Listening Skills Learning Station Card #3

# Make a stack of five (5) cubes. Each cube must be a different color.

Didax Inc.®

Listening skills Learning Station Card #4

# Make a stack of five (5) cubes. The same color cubes must not touch.

Didax Inc.®

Listening skills Learning Station Card #5

# Make a stack of five (5) cubes. Two colors must have two cubes each.

Didax Inc.®

Listening Skills Learning Station Card #6

# Make a stack of five (5) cubes. You must use at least two colors.

Didax Inc.®

Listening Skills Learning Station Card #7

# Blocks in Socks - 1

## Knowing the language

While many of the activities appear quite simple, the language required to understand the idea is relatively sophisticated. Many mathematical errors occur because students do not comprehend or misunderstand the wording involved.

When trying these activities with young students, one claimed to have "misunderheard" the instructions!

When giving specific instructions—such as those used in this activity—watch for students who "misunderhear." If the students give you a blank look, it is likely that they are unfamiliar with the language being used. Model the correct language. Write the words on the board for the students to see.

Examples of misunderstood words include:

pile     heap     collection

stack     join     connect

colors (see list page 10)

tall/short

Seriation (the concept of order in relation to quantity, especially size) words

compare        order

most/least       more/less

long/longer/longest

short/shorter/shortest

highest/lowest

---

***Without knowing it, students will develop addition and subtraction ideas.***

**1** **Blocks in socks**

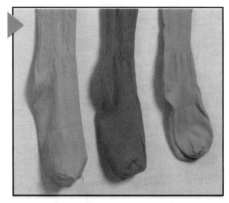

Collect a few brightly colored (clean!) socks.

Focus the students' attention on the collection of colored socks. Ask the students to name the colors.

Explain that you will be placing some cubes in each sock. Tell the students to close their eyes (no peeking!) while you fill the different socks.

State *"One of these socks has more cubes than any other."* Invite a student to come and feel the socks to see which one has the most cubes. Once the student has chosen a sock hold it up for the whole class to see. State *"Now let's see if this sock holds the most blocks."*

Empty each sock and stack the cubes on top.

Ask the students whether their classmate was correct. Why?

Encourage discussion and then the chant *"Most blocks in socks – (sock color)."* When you have discovered the least number of blocks in a sock *"Least blocks in socks – (sock color)."*

Continue the game until the students grasp the idea.

## ② Pass the sock

Students are seated in a circle, with a few cubes in front of them. Every fourth student has a sock.

Tell the students with the socks that you will be holding up a number of fingers. (Note: You can alter this activity by holding up a card with dots showing on it or cubes. Later you could use numeral cards or word cards.) The students with the socks in front of them will place that many cubes into the sock. When you call out "pass," the sock is passed to the person on the left. Repeat for the next two students. On the fourth move, the students take the cubes out and make a stack in front of them.

*"Who has the highest stack?"* Hopefully they are all the same. How many cubes altogether? Make a copy of the stack to stay on display.

Instruct the students to break-up the stack and put the cubes back in the sock and keep one or two in their hand and not tell anyone what they have. The sock is then passed to the next student. Remind the students to only take one or two cubes.

Some cubes will be left in the sock. Ask the students to take them out of the sock and stack them. Encourage the students to compare their stacks. *"Which student took the most cubes altogether?, Who took the least cubes altogether?, How many cubes are in the small stack?, How many cubes are in the larger stack?"*

## ③ Find the difference

If the students are ready, ask questions that involve difference—not an easy concept for young students. You do not have to "teach" how to solve these challenges, the students will do so naturally. For example,

Find the most blocks in socks

I put two blocks in the sock

I put three more blocks in the sock

I take three blocks from the sock

## Feely bag

Some teachers might call these "feely-bag" activities. Young students are fascinated with the hidden, so it can be played with a real sense of fun.

Originally, we did not see the challenging learning potential in the activity. As we worked with the students, we did no more than ask questions. With practice, they soon mastered the vocabulary involved, even the difficult difference idea. The students took over the idea and for days that is all they wanted to do: play Blocks in Socks! As quickly as they became involved, they abandoned the idea. It was as if something had told them, "You have had enough." And we think that was true. This game had provided a whole range of ideas in a new setting: that was the challenge. As avid learners, the students set about mastering the game, which they did. Then, it seemed, it was a case of now for the next challenge ...

Too many activities offered to students are so worked over that the students lose any interest they may have had. A good teacher knows when to move on because the students send out easily interpreted messages when enough becomes enough.

But, what about the student who has not mastered these ideas?

That's a fair question, but be assured that student will be much closer to mastering those ideas than before he/she met Blocks in Socks. Students develop and learn at different rates, and the concept may gel in the next series of activities they undertake.

# Blocks in Socks - 2

## Purpose

Students will become familiar with the terms *least* and *most* and will be able to order small amounts of cubes.

### ① Pin-up socks

Pin three socks securely to a board at a height that students can reach. In each sock is a number of cubes. Above the socks, pin words like **most**, **least**, or **smallest number**, etc.

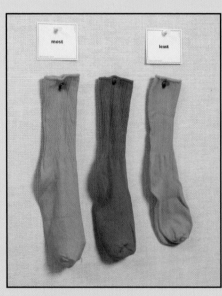

### ② Feel and match

Students feel the socks, guess and provide answers at a prearranged time. They match the words to the correct sock with a chant like, *"This blocks in socks has the most."*

## Guide to using this photocopiable resource

### Blocks in Socks Cards

#### Cards 1 to 4

When you first set up the challenge, make the variations obvious. As the students gain experience with the activity, the differences between the number of cubes may become subtler.

#### Cards 5 and 6

While some students may line up the socks from right to left, most will line them up in order left to right using the first attribute as the start.

#### Cards 7 and 8

One of the most absorbing activities at this level is:

Pin a card on the board, pass a number of socks to individual students who are then challenged to place a number of cubes in their sock, obeying the instruction card on the board.

**most**

Blocks in Socks Card #1

**least**

Blocks in Socks Card #2

**more**

Blocks in Socks Card #3

**less**

Blocks in Socks Card #4

**Order most to least**

Blocks in Socks Card #5

**Order least to most**

Blocks in Socks Card #6

**The number of blocks in socks is the same**

Blocks in Socks Card #7

**The number of blocks in socks is different**

Blocks in Socks Card #8

# Pattern

Observing, representing and investigating patterns is fundamental to the development of mathematical concepts.

*Students needs to be able to:*

- **recognize** *(observe)*
- **reproduce** *(copy)*
- **identify** *(explain, describe)*
- **extend** *(describe a rule, continue)*
- **create**
- **translate** *(from one mode to another)*
  *for example, color to words: red, white, blue, red, white, blue to snap, clap, tap, snap, clap, tap*

## When working with students, note whether they can:

- generalize from a few examples
- describe/explain a rule
- make their own pattern and describe the rule used for creating the pattern
- predict, make conjectures

# Patterns come in a variety of modes:

***auditory***— rhythm, beat

***number***— square numbers

***visual***— shape, color, size, thickness

***movement***— dance, skipping, moving parts

***patterns in nature***— spider webs, leaf patterns

***events in time***— night and day, seasons, mealtimes

***symbols***— ABAB, AABAAB, AABAACAABAAC

***writing patterns***— ↑↑↑↑↑↑↑↑, aaaaaaaa

***figural patterns***— numerical, spatial

# Pattern

*Science and mathematics are both trying to discover general patterns and relationships, and in sense they are part of the same endeavor. The history of science is replete with discoveries of patterns in nature. Cycles, geometric designs, cause and effect, physical laws, etc. are but a beginning of countless examples that could be listed.*

Rutherglen, Ahlgren
*Science for All Americans*, Oxford University Press

## The importance of pattern

A rich experience in pattern-making will allow students to see order in the number system and to make relevant connections with prior mathematical learning experiences.

The discovery of patterns is a lifelong characteristic of the thinking person. Without knowing it and beginning at birth (or possibly even before that), an individual uses the pattern-finding process to carry out all learning.

We call the process DSOP

It works like this:

1. Collect DATA: this is happening all the time through ALL the senses: sight, touch, hearing, taste and smell.

2. SORT (classify) this data: this is a constant function of the brain.

3. Find an ORDER in this data.

4. Now, discover a PATTERN.

This appears to be a natural function of the brain and the real mathematical growth occurs when the pattern is tested and tested and even tested again. When that **PATTERN** assists in the formation of a **HYPOTHESIS**, that is, it has been successfully tested, a sound mathematical process has taken place.

# Snap-Clap

## Purpose

Students develop a variety of patterns and represent these patterns in many ways.

Snap

Clap

Tap

We believe strongly in the developmental notion that individuals need a physical experience to create, strengthen and consolidate concepts in the brain. For this conceptual development to occur, it is paramount that a realistic concrete experience has occurred. We see concrete in terms of reality, rather than solid objects. In this case, the cubes provide a platform for a real developmental experience.

## *Rhythmic patterns create a deep physical experience.*

**1** Rhythmic hand clapping, finger snapping and even knee tapping provide wonderful opportunities for creating patterns in early childhood. Students will create their personal series as the teacher (or another student) taps a rhythm.

As each individual acts the pattern, the other students are invited to join in. Young students can follow the most sophisticated patterns set by their classmates.

Different taps can assume different actions—even the following sequence!

**snap, snap, snap, clap, snap, snap, snap, clap, snap, snap, snap, clap** could mean, "Now is the time to get your lunch."

snap  snap  snap  clap  snap  snap  snap  clap  snap  snap  snap  clap

**2** Create a train (a long stack) of two blocks of one-color and two blocks of another color. The first color represents SNAP, the second CLAP.

Read the train as:

**snap-snap, clap-clap, snap-snap, clap-clap, snap-snap, clap-clap, …**

Strike up a rhythm and the students can snap-clap according to the codes on the blocks. Keep the pattern going after the train ends.

snap  snap  clap  clap  snap  snap  clap  clap  snap  snap  clap  clap

Introduce new colors and, hence, new actions.

snap  snap  tap  clap  snap  snap  tap  clap  snap  snap  tap  clap

**3** In time, create written snap-claps. Label each color with a letter.

Hence this pattern becomes an **AABB** pattern.

*snap-clap pattern (AB pattern)*

snap clap snap clap snap clap snap clap

*snap-clap-tap pattern (ABC pattern)*

snap **clap** tap snap **clap** tap snap **clap** tap

*long train (ABCABC pattern)*

A   B   C   A   B   C   A   B   C   A   B   C   A   B   C

*longer train (AABBCC pattern)*

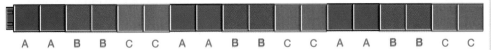

A   A   B   B   C   C   A   A   B   B   C   C   A   A   B   B   C   C

**4**   1.   Try a variety of labels to describe the moves.

2.   Use number digits to describe the pattern. Understand these are digits only and do not represent a quantity or order. (Yes! Young students can handle this very competently.)

**1, 1, 2, 2, 1, 1, 2, 2, 1, 1, 2, 2, 1, 1, 2, 2**

3.   Now, translate this "snap-clap" code.

2b, 2w, 2b, 2w ...

Two blue (or black), two white, two blue, two white and so on.

OR

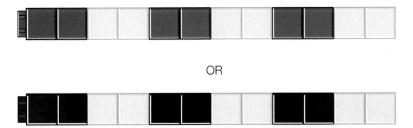

There are obvious links with other curriculum areas; for example, language, music and physical education.

## Make a display

Using the photocopiable resource on page 23, students may record their best trains.

These may serve as masters for other students to copy.

Make a collection of different trains for display.

## Learning from experience

One student set up this pattern.

In a less than thoughtful manner, I reacted, "Hey! You can't do that!" The drop of jaw and the defeated look on the student's face told me instantly I had made an error of judgment. In an effort to rescue the situation, I offered, "Best you tell me all about it." Happily, the student went ahead:

**snap, snap, clap, SILENCE, snap, snap, clap, SILENCE ...**

The student was thinking well ahead of me; silence was part of the pattern. The student had created a more sophisticated pattern—each pattern unit was made up of four elements: 1. snap, 2. snap, 3. clap, 4. silence.

It was a timely reminder that premature and ill-thought-out interjections can stifle the creative thought process.

# Making Long Trains

## Purpose

Students will realize that patterns can go on forever.

Patterns may be labeled and described in many ways.

*The pattern-building experience continues.*

This is a group activity.

Students will realize that a pattern can go on forever.

Ask the students to select two colors (contrasting colors such as blue and white are ideal) and stack them in an AB pattern. You can use any letters; by now students will recognize the pattern. Now, ask the students to join the stacks together to make a long, long train.

Meanwhile, other groups of students have done the same thing. It invokes a sense of wonder as the students compare the lengths of the trains. Make more sophisticated trains.

Notice that a repeat is included so students can see the start of the pattern. With experience, students will create more difficult patterns.

## Guide to using this photocopiable resource

- Students may color their patterns and cut out the pattern strips.

  (Teacher note: Blocks may be placed on the strips to model the pattern. We have noticed that young students prefer to color-in larger as compared to smaller squares.)

- Glue strips end-on to create really long trains.

- Make long strips that repeat a pattern several times.

- Create long strips made up of one long paper strip and one short paper strip.

- Build patterned paper strips to match the length of an object or encircle an object such as a drink bottle or a picture frame.

- Challenge students to devise descriptive labels for the patterns they have created. The label will enable others to recognize the pattern.

Note: The strips are smaller than the actual cubes. Enlarge at 125% to make them the same size.

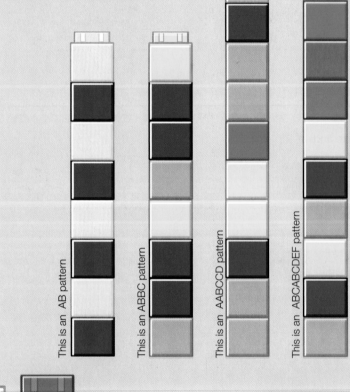

This is an AB pattern

This is an ABBC pattern

This is an AABCCD pattern

This is an ABCABCDEF pattern

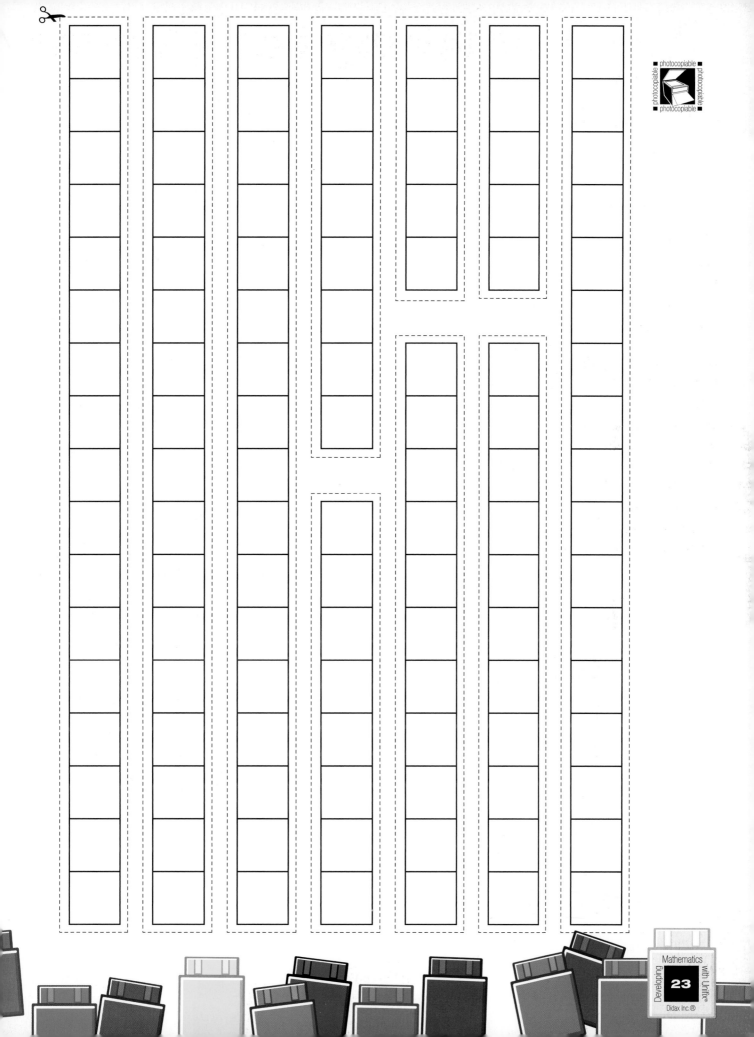

# Repeating Patterns

## Developing ways of describing patterns

There is no doubt that students are developing their mental, physical and social skills by their interpretation of the patterns they discover in their environment. This process is natural.

Having made the mathematical connection, the discovery is communicated by a very special language, which is the mathematics so frequently prescribed for school lessons. Too many students (and adults) are confused by this special language because they have done too little of the discovery process so vital in developing the mental connections of mathematical competency.

In short, during the early part of a student's life, the teacher will provide a very sound foundation by concentrating heavily on developing pattern concepts accompanied by the most appropriate vocabulary (language).

## *Make a collection of different patterns to allow students to discuss ideas.*

### ❶ Top 'n' bottom

Arrange a number of cubes in a top 'n' bottom pattern.

List the keywords and phrases used by the students when they describe patterns.

### ❷ Colored tops 'n' bottoms

Use two (later on, more) colors to make an interesting pattern.

### ❸ Ones, twos 'n' threes

 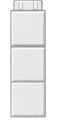   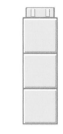

Make a line pattern using one cube, two-cubed stack and/or three-cubed stack.

### ❹ Repeat-a-design patterns

Students create a shape and repeat that shape to create a pattern. The pattern may be horizontal or vertical. Encourage the students to invent more repeating patterns.

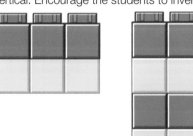 

Ask students to continue these patterns.

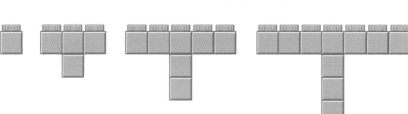

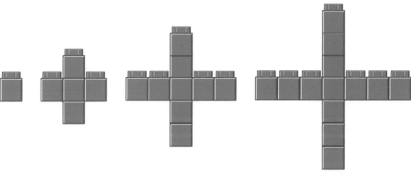

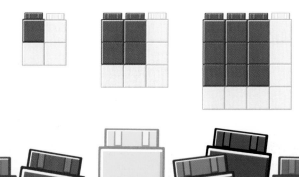

## Surrounded by pattern

Consider the notes in your favorite song, the pattern of the tiles in your shower, the flow of traffic at around 8:30 a.m. and the colorful clothing patterns worn by young and old. Have you thought about it? Mathematics is the study of both visual and numerical patterns. In fact, mathematics can be found all around—you are surrounded by mathematics.

Students need to be encouraged to observe patterns in order to be able to predict how the pattern will continue. This is a very good example of DSOP (see page 19) in practice.

# Growing Patterns

## Purpose

Students create a variety of repeating patterns.

### *Not all patterns repeat in the same way.*

As you can see by the set of illustrations, there are no set rules for these patterns. All the patterns below begin with Starter Card #1.

## Growing patterns

The secret to this type of pattern creation is to allow the students to recognize another form of pattern-making. In turn, this will encourage the creation of and search for patterns in the environment.

When creating your own template, make the squares fractionally larger than the actual cubes; this will allow the students to see the outlines.

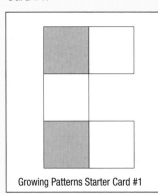

Growing Patterns Starter Card #1

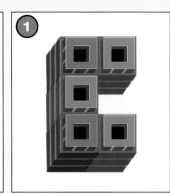

1

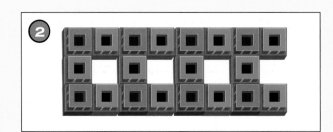

2

## Guide to using this photocopiable resource

Use the challenges opposite to create Learning Stations where students respond to the challenge on the card.

## Growing Patterns Cards

Cover the cubes on the *Growing Patterns Starter Card*. Of course, colors of the cubes will vary, but do not insist on the same set of colors. (In time, the students will realize the need for consistent colors.) Add cubes (of the same design) to the original design. The diagrams on this page will serve as a good guide.

At first, some students may find it difficult to spot the initial design. But, with experience and maturity, students will be able to create growing patterns and recognize the original pattern in other models.

Laminate the starter cards so they can be used in the Learning Stations. Encourage the students to plan their own starter cards.

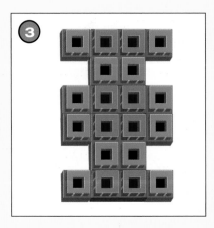

3

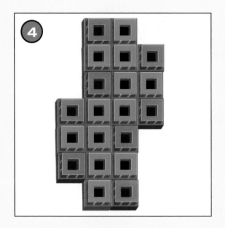

4

To allow students to move their creations, limit the size of the model. Often, this model will not be completed in one session.

Students will create their own beginning shape or pattern and keep it a secret. When the pattern is finished, other students are challenged to discover the original design. This is not always easy!

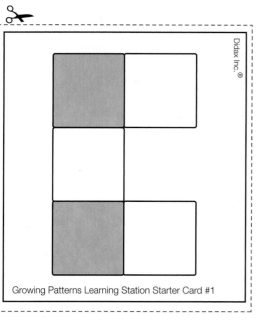

Growing Patterns Learning Station Starter Card #1

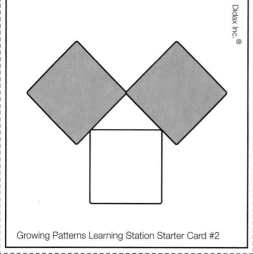

Growing Patterns Learning Station Starter Card #2

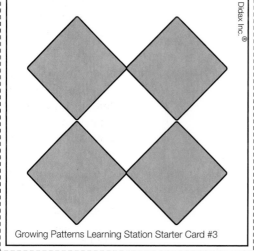

Growing Patterns Learning Station Starter Card #3

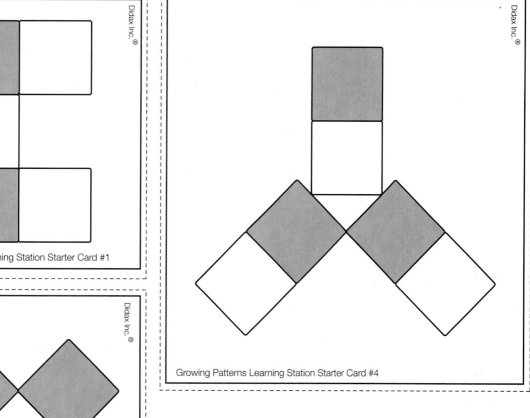

Growing Patterns Learning Station Starter Card #4

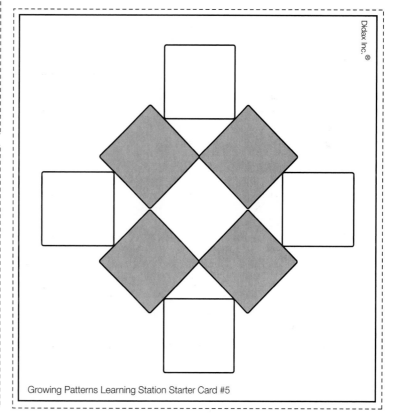

Growing Patterns Learning Station Starter Card #5

Didax Inc.®

# Cars in the Train

# Barn – 1

## Purpose

Students will describe patterns.

Students will look for relationships between patterns.

## Developing the ability to recognize patterns

This activity needs to be played many times to allow the students to assimilate the various patterns. You will see the benefit of this practice when you begin to develop counting and number activities.

### Encouraging explanation

Take photos for frequent discussion sessions. You will be amazed how competent the students become at recognizing various sequences.

## Play trains to create an involved pattern discovery journey.

Instruct the students to make an AB train with green and yellow cubes. Define the green cube as being represented by the letter A and the yellow cube as B. Refer to a single cube as a car.

Check the students' understanding by asking *"What's a car?"* Make sure the students understand this definition. (A car is one cube.)

A   B   A   B   A   B   A   B   A   B   A   B

Ask the student to make an AB train. Once a long train has been created, explain to the students that *"We are going to put the cars into barns. The first barn is a one-car barn. Let us put some cars in the barn. Watch how it is done."*

### One-car barn

Note the rules for parking in the barn.

1. Each time, the first car in a barn is A; in this case, a green cube.

2. Its joiner will point to the left; see the 2-car barn for the next example.

3. For the pattern to become obvious, about six layers need to be placed in each barn.

### Two-car barn

Place the train in a two car barn. Remember to follow the rules and that two cubes means two cars. Begin with an AB joined together.

Ask the students *"What do you notice about the cars?"* Students will observe that the same color is in a line.

*Did this happen in the one-car barn? No! Why? Let's try a three-car barn.*

Continue exploring the different-sized barns.

### Three-car barn

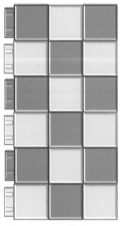

Note the checkered pattern.

### Four-car barn

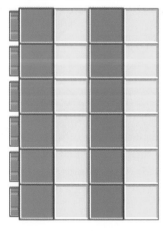

How is this related to the two-car barn?

# Barn patterns

This is an A B B C A B B C pattern. How will it look on different grids? Hint: Look at the lines on the models. In these activities students will recognize:

### checker-board

### (to the) right diagonal line

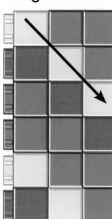

### straight line

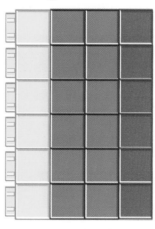

## (to the) left diagonal line

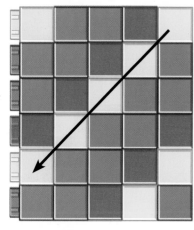

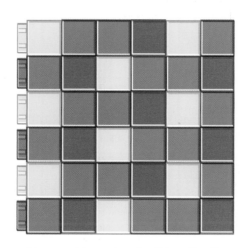

How would you describe this pattern?

## Using letter codes

We can letter-code these arrangements. For example; using an ABB pattern, cars in the four-car barn will look like this:

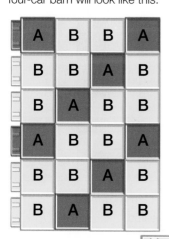

**or**

ABBA

BBAB

BABB

ABBA

BBAB

BABB

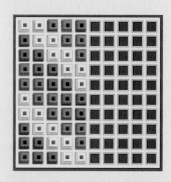

# Cars in the Train
# Barn – 2

**D T E S**

**Purpose**

Students will describe patterns.

Students will look for relationships between patterns.

### Predicting Patterns

As students become engaged in the search for patterns they will start to predict what will happen next.

In the ABCABC train pattern, visual patterns will repeat in the 3, 6, 9, 12... barns. Some students will notice this pattern. Others will start to predict what will happen one before and one after a multiple of three. Once the initial pattern changes (from ABC) to say ABCDABCD, then the whole new discover process will begin.

*The journey continues.*

### Numbers and cars

This is an ABC pattern.

When we make a long train and then place the cars in barns, the arrangements of the cars look different but they will show the same pattern. Consider these barns if you change the pattern to a 123 pattern. What would you expect? Traditional counting by threes is evident in the 3 and 6 barn – why?

| 1 | 2 | 3 | 1 | 2 | 3 | 1 | 2 | 3 | 1 | 2 | 3 |

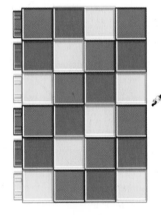

Note: The ABC pattern repeats in a cycle of three.

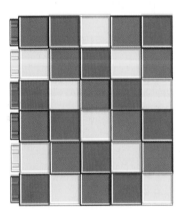
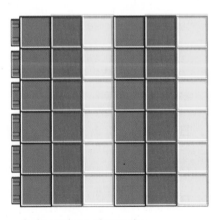

A six-car barn shows the cycle repeated.

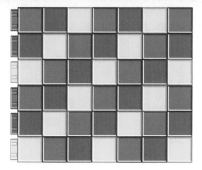

How is the seven-car barn related to the four-car barn?

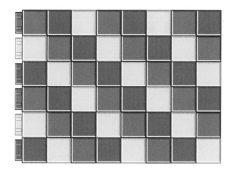

How is the eight-car barn related to the five-car barn?

## Hints and tips for using Cars in the Train Barn activities

- Young students will have no problems creating the Classroom Railway Yards. Keep your digital camera handy because the results are quite spectacular and students will enjoy revisiting their patterns.

- Use your discretion when you introduce the grids. Students will need considerable language experience before attempting to color in the grids on page 35.

- Treat the larger grids as extensions of the students' experiences.

- Students may color the grids and challenge the class to discuss the basic pattern; for example, is it an ABC pattern or a AABB pattern?

- When appropriately colored, these grids will make a fine display.

- It is important that the cubes be joined in a consistent manner, otherwise students may not form the patterns correctly. This will impact on the developed final product.

- As students become experienced with the Railway pattern-making, you can expand their vision. Ask them to put the cars into different barns. Cards may be used to indicate the length of each car. Ask the students to break the train into cars, beginning with a two-car barn and continuing with other barns.

- Color the grids using different combinations of colors. Do different color combinations make the pattern easier to see?

- Coloring the various squares can be a challenge for students. A mistake can be easily corrected.

## Guide to using photocopiable resources on pages 32–35

### The Railway Yards (pg. 32)

Continuing the spirit of the exploration of pattern, students can label the barns in the railway yards. You can use a label as a challenge by instructing, "Park an ABC pattern train in this barn."

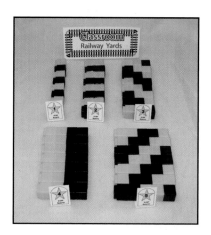

### Cars in Barns Practice Sheet (pg. 33)

Use this sheet to encourage students to place the cars correctly. For the sake of understanding the pattern, it is important that the shoulder end of the cube is placed consistently and the arrangement begins with the same color each time.

### Blank 1 to 10 Grids (pg. 35)

Students may use these plans to help plot the layout of the barns. Also, the grids can be colored and used as challenge cards for others in the group. Try making an AAB pattern on each grid.

# Classroom
## Railway Yards

fold to make the sign stand

**1**
car
barn

fold

**2**
car
barn

fold

**3**
car
barn

fold

**4**
car
barn

fold

**5**
car
barn

fold

**6**
car
barn

fold

**7**
car
barn

fold

**8**
car
barn

fold

**9**
car
barn

fold

**10**
car
barn

fold

**11**
car
barn

fold

**12**
car
barn

fold

Cars in Barns Practice Sheet — Please enlarge at 125% to make actual size.

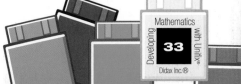

# ABCD Pattern on 1 to 10 Grids

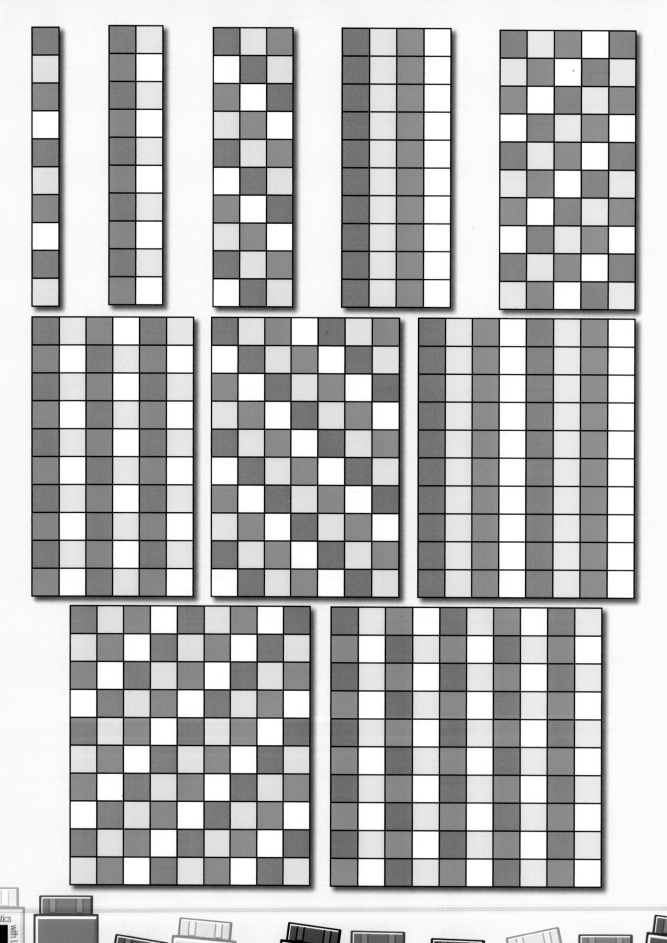

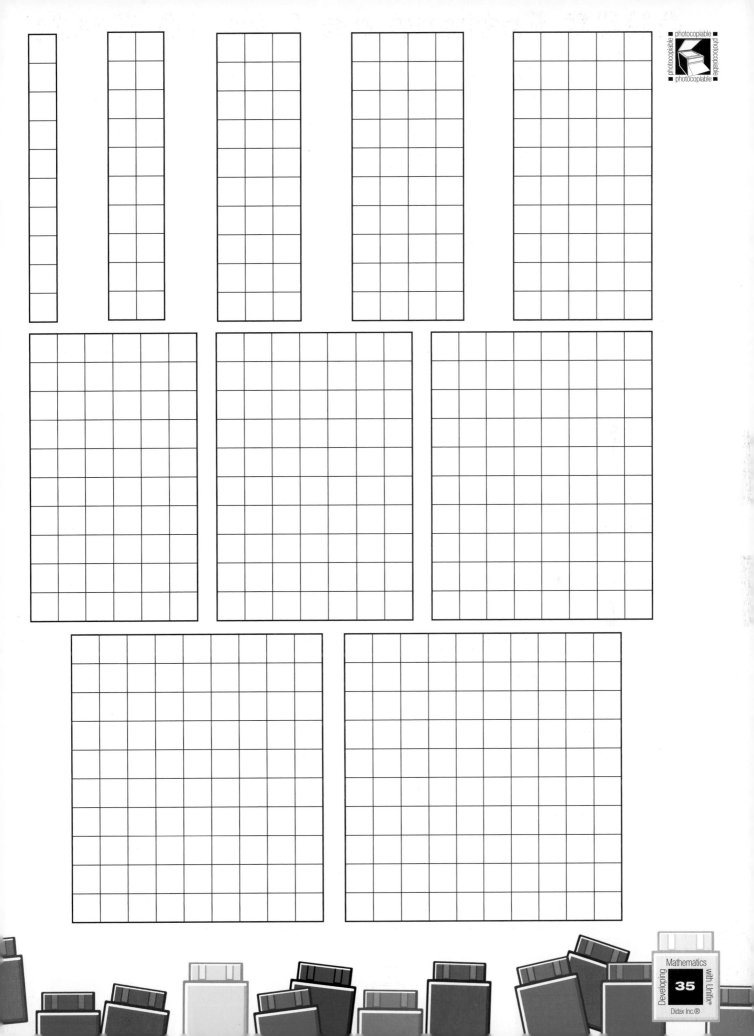

# Cube Stacks

## Purpose

Students will use logic and patterns to solve problems.

Students will prove by exhausting all possibilities.

## The importance of pattern

There is interesting research in these combination activities. Younger students may explore the patterns of the various combinations while older students might see if they can devise rules for the number of combinations of any group of cubes.

***Avoid telling the students formulas for solving these challenges. If they find one that works, celebrate.***

At this stage of development and experience the investigation, discussion, tentative hypothesis, and the testing of that hypothesis are far more important than the search for an answer.

Note: The challenges multiply rapidly.

## Endless investigation possibilities.

 Give the students the following instructions.

*Using two cubes of the same color each time, how many different stacks can you make?*

*If you use a red cube and a blue cube, will you be able to make more different stacks? Why? Does the order of the colors make it a different stack? Why?*

Allow the students to talk this through. Do not rush the response but in time accept a consensus that when we are looking at the stacks in the same way each time, a different order of color will mean a different stack.

**Teacher:** Remember when we stacked two cubes, one red and one blue? How many different stacks did we make with those cubes?

**Students:** There were only two stacks.

**Teacher:** Very good thinking. Now, stack three cubes this time—let's use red, white and green—in as many different ways as possible.

How many did you create?

**Students:** I think there are five.

**Teacher:** Do you agree, class?

**Students:** No, I think there are six different stacks.

Keep questioning the students and sorting the stacks. It will not be long before there is agreement that there are six different stacks.

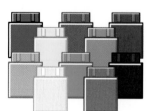

*This time use ANY two colors. Make all the stacks possible. When all the stacks have been made, sort them, then ask a friend to tell you how the stacks are sorted. Here are a few stacks.*

**2** Try these stack-the-cubes-by-color challenges.

Three cubes

- Use three cubes and two colors

Four cubes

- Use four cubes and one color

- Use four cubes and two colors

- Use four cubes and three colors

- Use four cubes and four colors

- Use two pairs of one color and two pairs of another color

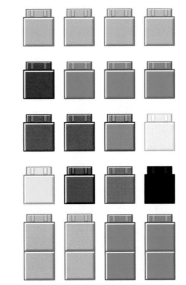

Compare each collection of stacks. (There are too many stacks when you set a similar series of challenges with five cubes.)

**3** The illustration shows all the stacks you can make with four colors. Find any patterns which could give you a rule so that you can tell how many different stacks there will be with any number of different colored cubes.

**4** Challenges

- Try four different-colored cubes! (24 stacks)

- What about five different-colored cubes? (120 stacks)

- If you tried six different-colored cubes each time, how long would it take if you could stack one per minute? (720 minutes)

- Imagine 10 cubes … WOW!! (3,628,800 minutes—a long time)

Investigate the wonderful book, ***Anno's Mysterious Multiplying Jar***, Mitsumas and Masaichiro Anno, Bodley Head, London, 1982

## Purpose

Students recognize patterns.

Students create patterns according to set rules.

## From pattern to number

The snake patterns lay the foundation for number patterns, hence counting patterns, which lead to multiplication patterns. Do not rush the students to formal presentation of these ideas: remember we are working mainly at the D and T stage of DTES.

You will be pleased to discover how well the students can count. These activities encourage students to count in 1s, 2s, 3s, 4s, etc., eliminating the need for more formal and frequently boring counting activities. Experience has shown that students are doing what comes naturally.

### Learning from each other

Students learn a lot from each other, so encourage these activities to be played in groups, rather than with the whole class.

### What to look for

You will know that students are mastering this skill when they attempt to make snakes of larger numbers. Later on, the snake patterns can be transferred to the operational grid and container (see page 89) and the students will have a new set of patterns to consider. As they strengthen these skills, different grids, for example, 8 x 8 or 9 x 9, will be presented in different reference frames.

# Snakes

## Children develop the concept of groups naturally.

This is an activity which can involve all students at a very early age. Take careful notice of the types of patterns the students will build; you may be very surprised.

A snake is similar to a train in that it is a series of cubes stacked end-to-end. But, the secret to a snake is the naming of the pattern.

For example:

This is a 2-colored-1 snake.

This is a 2-colored-2 snake.

Ask the students to build as many different 2-colored-2 snakes as possible.

Show the two snakes depicted below and ask "*Are these the same snakes?*"

Some students will argue yes, others no. At this stage, allow the students to decide.

Give the snake a name: 2-colored-2 snake.

Build a few more snakes: use different colored cubes each time to establish the naming convention.

Build some others; for example,

a 3-colored-2 snake

a 2-colored-3 snake

a 3-colored-4 snake

Ask the students to make a collection of snakes. Each snake must have twelve cubes.

> *Create a Snake-Pit somewhere in the classroom. You could use a shallow wicker basket or a strip of cloth.*

**1** **Snake-Pit games**

Place a number of twelve cube snakes in the pit. When a snake is called—for example, 3-colored-4 snake—two (maybe three) students run to the pit to find the appropriate snake. First to return with the snake wins! It will not be long before the students control this game by themselves.

Make a snake-pit with 10 to 12 different snakes. The snakes may have different patterns and/or different colors. Play the same challenge.

**2** Decide the length of a snake; say, nine cubes. Create all the snake patterns you can.

3-colored-3 snake

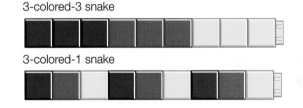

3-colored-1 snake

**3** A snake must have twenty-four cubes. Make as many different snakes as possible. Here are a few of the possible snakes. You could build a 2-colored-2-snake, a 3-colored-3-snake, a 3-colored-4-snake and a 6-colored-2-snake.

Remember: Each snake must show a **pattern**.

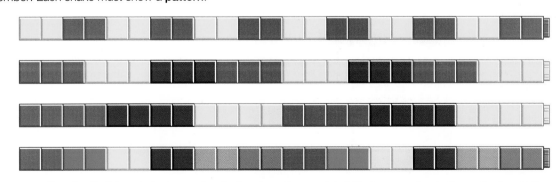

By now, you will realize that the students are developing a most powerful ability to recognize patterns, the key skill needed to see order and to make connections in mathematics. You will see excited students searching for patterns, using materials to create patterns and developing techniques to record them.

# Unifix® Pattern Building Underlay Cards

*A more controlled pattern-making experience is provided when using these materials.*

Operational grid and container with cubes and pattern building underlay cards.

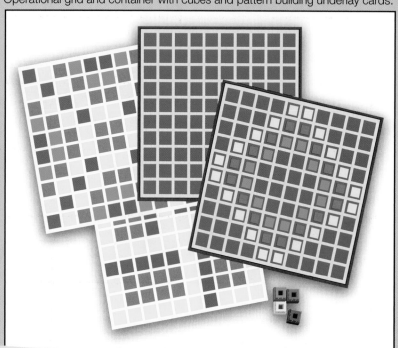

*Building an ABCD pattern on the operational grid.*

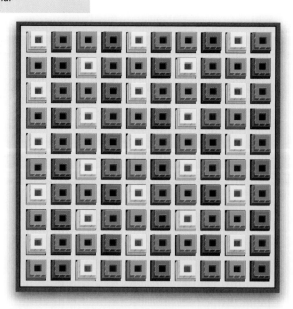

## Guide to using this photocopiable resource

### Design on a 10 X 10 grid

Our experience has shown that students prefer to use the Unifix grid rather than a photocopied grid sheet. Of course, if you do not have access to the Unifix grid, the paper (or card) sheet will be suitable.

*Allow time*

The creation of a pattern on a 10 x 10 grid is a challenge which requires a dedicated effort to complete. We recommend no more than 10 students working in pairs on this project at any one time. Make sure you have a secure storage site because no student will wish to break down his/her pattern before it has been shown to classmates.

Some students may wish to color in a grid as a record of the created pattern.

*Keep a record*

Make sure you take a photo of the results. Not only will this be a valuable item to include in a student's portfolio, but it will provide a wonderful reminder of a most worthwhile activity.

*Hint:*

Create the same pattern but use different color combinations. Which one looks the best?

This grid will need to be enlarged 120%, so the cells are 21 mm square. This allows the students to see part of the line around each cube.

# Number Patterns

## Experiencing a variety of number patterns

From practice with young students, we realize that if students understand the principles of pattern they can represent number patterns in a variety of styles.

Notice, for example, that the triangular number pattern may be represented in more than one form.

## *Relating pattern work to number work.*

Encourage the students to create a variety of number-related patterns.

**1** Montessori number patterns

**2** Domino number patterns

**3** Threes number patterns

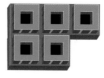 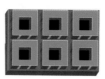 

**4** Other **figure patterns** may include: polygon patterns; for example; triangle, square, hexagon, etc...

triangle

triangle

square

**5** This is one **polygon** number pattern.
Find others.

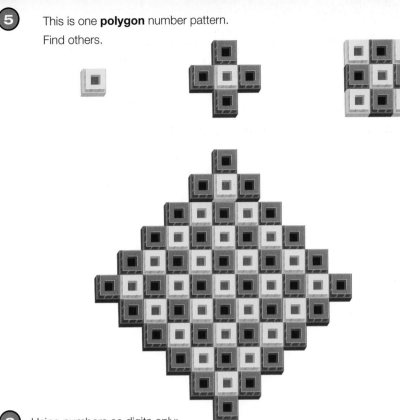

**6** Using numbers as digits only:

1, 2, 3, 1, 2, 3, 1, 2, 3, …

Make these number digit patterns.

- 112233, 112233, …
- 1234, 4321, 1234, 4321, …
- 456, 456, 456, …

These number-counting patterns may be matched with appropriate cube stacks.

- 2, 4, 6, ___, ___, ___
- 10, 8, 6, ___, ___, ___
- 1, 3, ___, ___, ___, ___, ___
- 12, 9, ___, ___

**7** Counting patterns

Counting the snakes (see pages 38 and 39).
Counting by 2, 3, 4, etc.
In the example above, students could be encouraged to count by threes.
Encourage students to count a collection of Unifix Cubes by using their pattern-making skills. Counting tells how many, but what is the best way to find how many?

Counting the snakes (see pages 38 and 39).

## Transfer these ideas

Up to this stage, we have used the Unifix® Cubes to develop pattern ideas in mathematics. Use other materials to show that students have developed and mastered the discovery and recognition of patterns.

Collect various patterned objects—wrapping paper, fabric and wallpaper—to discover patterns. Challenge the students to name a type of pattern. Allow them to create their own labels.

Various tiles, plastic and ceramic, make ideal pattern-search/creation materials.

While pattern making and recognition is a natural function of the human mind, students need constant practice and language development in order to make their discoveries and understandings easily communicable—a vital aspect of the mathematical learning process.

# Counting, Place value & Operations

120

145

5  250

30

Understanding place value is important to achieving good number sense, estimating and mental math skills and to an understanding of multidigit operations.

*National Council of Teachers of Mathematics*

# Counting Number and Place Value

The development of number skills does not rely simply on counting and number facts. Fundamental language usage, such as working on concepts of **more or less** and the ideas involved in **difference and same**, is extremely important at this stage of a student's mathematical development. The use of Unifix® Cubes and support materials will assist in overcoming many of these shortcomings. These concepts include:

## Counting

Most people on earth can count. In fact, we believe that the structure of our brain promotes humans to do this very basic activity. But we do not use the same words to express this counting ability. So, counting happens before words are applied. It is with the application of the correct words to the right concept that the teacher has such an important role to play. Counting is much, much more than a nursery rhyme.

## Place value

The concept of place value was introduced to Western civilization in the thirteenth century. It means that a digit has a value according to the position it holds in a number. So, in 444, each digit has a different meaning. The concept is not exclusive to our base ten system.

Do not assume that just because a student can identify which digit is in the tens, hundreds, or ones column and can tell you 862 is made up of 800 + 60 + 2, he/she understands place value. When asked how many tens in 862 the same student may not recognize there are 86 tens.

## Ordinal number

A counting number used to indicate a definite place in an ordered sequence of objects. Example: 1, 2, 3, are called first, second and third.

## Cardinal number

A cardinal number indicates the numbers of items. There are 4 items in this group: (● ● ● ●). Also, it is known that $1 < 2 < 3 < 4$, and so on.

## Understanding of the relationship between the four operations

Understanding the connections between addition, subtraction, multiplication and division is fundamental to developing sound number skills.

Addition and subtraction are the inverse of each other. For example, the inverse of adding 5 to a number is subtracting 5 from the number. Likewise multiplication and division are related: $3 \times 8$ is 24, $24 \div 8$ is 3.

Division is the successive subtraction of equal groups and multiplication is the successive addition of equal groups.

## The acquisition of mathematical concepts is very natural

The abstract presentation of our number system with its ingenious place value system is dumped on our young students far too early in their developing formal mathematical life. As we prepare to present a wide range of experiences for young students, let's accept a few very basic understandings.

There is a huge difference between knowing a concept and presenting it in a formal fashion, such as writing numbers and presenting equations.

We tend to teach the body of knowledge rather than provide the experiences out of which a body of knowledge emerges.

*If students are to have a good understanding of numbers, it is extremely important that their understanding of place value should be thoroughly built up and should deepen throughout the elementary years, gradually extending and generalizing towards the millions and incorporating decimal numbers.*

Hilary Shuard, *Primary Mathematics Today and Tomorrow*, 1986

Consider the role that language plays in the thinking process. At the same time, realize that a person does not necessarily need to have language facility in order to think. But, for many students, talking it out is thinking it out. A student's language is one of the keys to his/her learning.

- Does a teacher need to be aware of all that a student talks about?

- Students can carry out worthwhile conversations in active classrooms; the rest of the students do not have to be quiet while one student talks it out with the teacher or anyone else, for that matter.

- Watch young students talk-in a new idea. Watch their faces as they configure an idea mentally. Watch them tentatively test out that idea.

# Three Cubes on Your Fingers

### Purpose

Any number, in this case three, can be represented in many different ways.

## Developing a picture of three

Within this activity is an amazing idea for young students to grasp: every picture of three is different, yet every one of them can be called three and, later, symbolized by 3.

The finger game does not work well when four cubes are used, but adaptations of the idea may be created.

At this stage we have used only language to share ideas. Now is the time to move to the E stage of DTES; E means explain and this can mean "students write, draw, or model the idea any way they wish."

This is an important stage before we move on to formally introducing the digits. If the students write formal responses do not be concerned. Check to make sure the student knows what he/she is doing, and, if he/she does, celebrate with the student, not the class. One of the problems is the tendency for students to copy another student's ideas. We need to check to make sure the student knows what is happening, but never challenge a student directly for copying.

## What is hidden in the digit "3"?

You have completed quite a lot of informal counting with some of the pattern-making activities. At this stage, all of the counting activity will be oral; the symbolic digits will be introduced shortly.

## Playing with threes

Working with the whole group, count out three cubes of the same color.

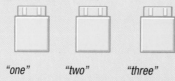

*"one"*     *"two"*     *"three"*

*"I have a group of three cubes."*

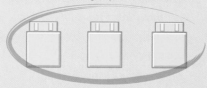

*Here is a group of three.*

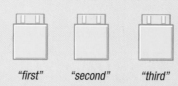

*"first"*     *"second"*     *"third"*

*"Show me the second cube. Now show me the first cube. Very good."*

You may observe that some students begin to count from the right side of the arrangement. Is that a problem? Not as long as the student is aware that he/she is counting in a definite direction. Students typically count a line of objects from left to right.

Count out three different colored cubes with the class. Ask the students *"Are any of the cubes the same?" "No."*

Now count out three cubes, with two cubes the same color. Ask the students:
*"How many cubes altogether?"*
*"Three."*
*"How many cubes are the same?"*
*"Two."*

Make two groups of three cubes and ask the class *"How many cubes altogether?"* Repeat this challenge with more cubes.

## Guide to using this photocopiable resource

Make a collection of all the different "threes on fingertips" using the photocopy master on the facing page.

## Three cubes on your fingers

Arrange the students in groups of four to six. Ensure that each group has an adequate pile of cubes. Instruct the students to find three cubes of the same color then arrange them on the fingers of one hand so that no one else in the group has the same arrangement.

The students will take time to sort themselves out. Move from group to group offering help and advice to solve the problem, not answer it. Here are a few possibilities.

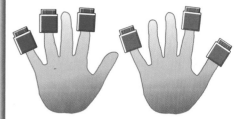

How would you respond if a student showed one of these solutions?

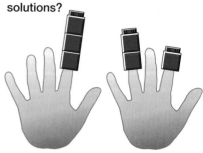

I hope you would challenge the student; just as you would any other: *"Tell me about yours, please."*

*"How many 'threes on fingertips' would there be if color mattered?"*

*"Lots!"*

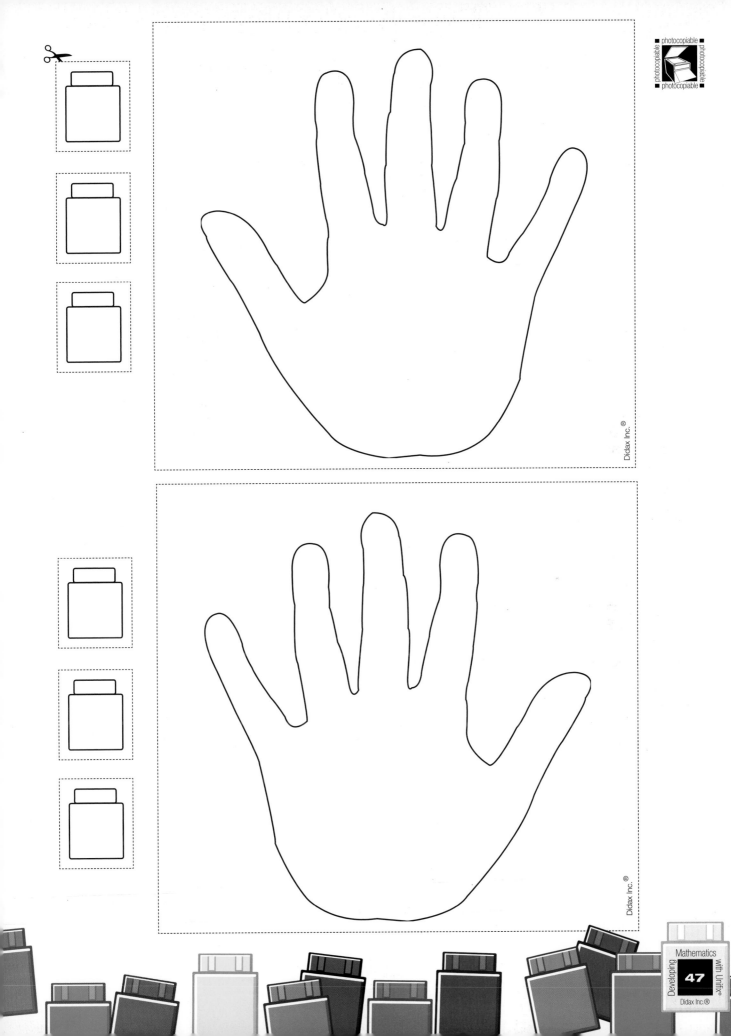

Didax Inc.®

Didax Inc.®

# Introducing Number Indicators

## Purpose

Students match a numeral to a quantity.

### Number indicators

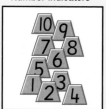

## Introducing numerals

This is the first time that numerals have been introduced—the S stage of D T E S.

Students thus far should

- count verbally,
- have a sense of one-to-one correspondence,
- appreciate the conservation of number; that is, the ability to see that six objects still amount to six, no matter how they are arranged.

In most classrooms, a number chart will be hanging on the wall and various number flashcards will be available. At this stage, we prefer to work on a number line because this seems to support the continuity concept of counting more appropriately than number grids.

## *Linking numerals with groups of cubes.*

Pour the number indicators on the table and show your outstretched fingers.

*Let's count my fingers with the correct numbers, starting with my thumb.*

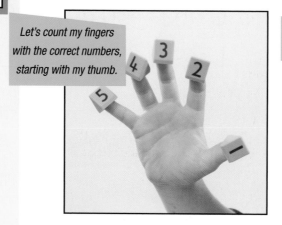

Students place the appropriate number on the counting finger.

*Count your fingers.*

  Show one cube. Place the correct number indicator on top of that cube.

  Stack two cubes and place the appropriate number indicator on the top of the stack.

Encourage students to experiment with various arrangements of cubes and number indicators. We could make a "story":

*"I have two cubes and I found two more. Now I have four cubes altogether."*

Students will make these stories naturally.

# Introducing Value Boats

## Finding order in the number system.

This is an exercise in seriation—placing things in order of magnitude. The attribute used with value boats is length created by the addition of cubes. We use the word "stacking" to describe the process.

Value boats provide another means of reinforcing the vital relationships between the abstract digit and what it represents.

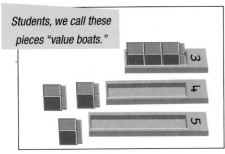

*Students, we call these pieces "value boats."*

**1** Show the set of value boats and scatter them on the floor or table. Each boat will hold only a certain number of cubes. Demonstrate by loading one of the boats with cubes, noting that they need to be stacked.

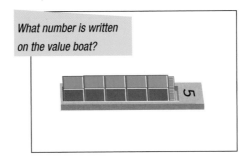

*What number is written on the value boat?*

**2** "What number is written on the value boat?"

"Five!"

"How many cubes are stacked in the boat?"

"That's correct. Five! The number indicator matches the number of cubes in the boat."

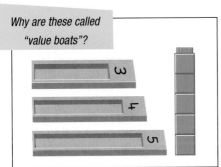

*Why are these called "value boats"?*

**3** Challenge students to stack any number of cubes, but no more than 10.

Find value boats to match the stack and encourage discussion about the word value.

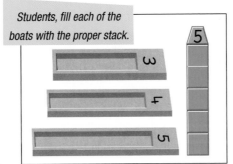

*Students, fill each of the boats with the proper stack.*

**4** Students fill each of the boats with the correct stack. The stacks may be taken out of the boats and appropriate number indicators placed on the top of the stack.

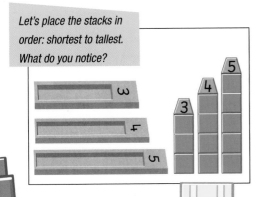

*Let's place the stacks in order: shortest to tallest. What do you notice?*

**5** "Let's place the stacks in order: shortest to tallest. What do you notice?"

Hopefully a student will announce, "We are counting to ten."

## Purpose

Students will match numbers to collections of objects.

Students will order numbers.

## Unifix® materials

*Value boats*

*Number indicators*

### From experience

While it is not essential to have these materials available, it is useful to be aware of them. It is not difficult to arrange similar activities to those promoted by these pieces.

### Teacher tip

In some older, established schools these pieces of equipment may be found in a storeroom. Clean them in warm, soapy water and use them appropriately.

These pieces consolidate the outcomes of all the teaching that has been taken place already.

Observing students using these pieces will help you assess how well they have grasped the concepts.

Place these items in a Learning Station where students can experiment with them.

# Make a Fleet of Boats

## Purpose

To develop students' counting and matching of a numeral to a quantity of cubes.

## More than counting

It takes time and experience for students to relate the name of a numeral to a corresponding set of materials. Just because a student can recite the numbers to 10, does not necessarily mean that he/she knows the set (or group) for each number.

As the students develop their skills:

- they will confidently (and accurately) identify small groups of cubes up to four or five,

- they will be able to count a set of cubes in any order,

- they will transfer their knowledge of a group of cubes—for example, threes—and use that knowledge to count a larger set of cubes,

- they will confidently imagine and consider numbers much larger than 10.

Frequently, we have observed students using various concepts of the four operations while they are developing these counting skills. For example, when a student notes that there are four groups of three because "I counted four lots of three cubes," a fundamental idea of multiplication is evolving. Likewise, through their counting experiences students strengthen their ideas of more and less.

Simply: Counting is far more than counting!

### *Apply knowledge of seriation in a challenging strategy game.*
### You will need:

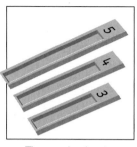

Three value boats per student

A collection of Unifix® Cubes

A 1–6 number die *(preferably)*

**OR**

A 1–6 dot die *(depending on student's experience)*

## Aim

To be the first to fill all of your value boats so that they become a fleet.

## Method

The cubes and value boats are piled on the table and each student randomly selects three value boats.

The die is thrown to show a number between 1 and 6. Each student selects that number of cubes and places them in his/her value boats.

Continue to roll the die until one student fills all of his/her value boats and wins that round.

On the next round, observe how the students select their value boats.

### Variation 1

Roll the die a sufficient number of times for the students to collect enough cubes to stack in three different stacks.

Taking turns, students select a value boat to match their stacks.

After each selection, students may rearrange their stacks to match an available value boat.

Points are gained for each successful match.

### Variation 2

Each student has a full set of value boats.

Taking turns, students roll a 1-10 die.

After each roll of the die, the students collect the matching number of cubes and place them in their value boats.

The aim is to fill all of their value boats.

Consider: Will the students learn anything if on each throw a different color cube is used?

# Order the Boats

## Purpose

Students will order numbers from 1 to 10.

***To play this game, students will rely on many basic mathematical skills.***

## You will need

One set of value boats per group

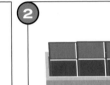

One set of number indicators per group

## Aim

This is a very challenging game, involving skills of ordering, quantity–number relationships and counting.

The aim is to arrange the pieces in order, 1 to 10. How it is done is the group's decision.

Play this game with a group of five to ten students at a table or in a circle on the floor. The number indicators are in a container.

## Working with numbers

By using these materials, the students are developing a strong understanding of the relationship between the number (numeral) and the *how many* idea of that number.

As well, by ordering the various groupings the *how many* concept is being reinforced by demonstrating the relationships between the various models. For example, which stack has more cubes? Is the seven stack the longest in the collection? In these activities, we are developing a sense of quantity. But we need to be careful with the words we use.

*longest* links with *length*

*widest* links with *width*

*highest* links with *height*

Then the comparative words will be included in the vocabulary development.

long, longer, longest
wide, wider, widest
high, higher, highest
small, smaller, smallest
big, bigger, biggest

Consider:
*I have a stack of three cubes.*

- *Double the length.*
- *Make it twice as long.*
- *Make it two times longer.*

All these say the same message.

## Method

**1**

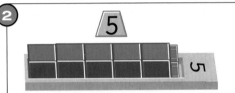

The first player takes a number indicator and selects the appropriate boat.

**2**

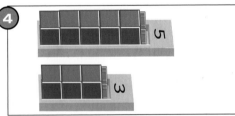

The student then fills the boat with the correct number of cubes.

**3**

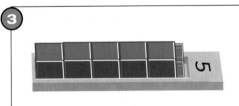

If the student selects correctly, the boat is placed carefully in the center of the group. If the incorrect solution is made, the student misses the turn.

**4**

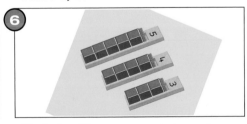

The next player carries out the routine and places the boat where he/she thinks it should be in relation to order of the other boat. Play continues …

**5**

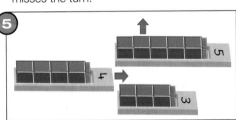

When there is no room to place a piece, the students must re-arrange the pieces to allow this one to be placed correctly. But they must move only a minimum number of pieces. The moves made are counted and at the end of the game are totalled. The fewer number of moves, the better the game.

**6**

Students will become game-wise and spread the pieces out, so the teacher will then limit the amount of space in which the pieces are to be placed!

# Measurement Sticks

## Developing measurement ideas

Students need a vast collective experience before they formalize the concepts connected with measurement. To be meaningful, the idea of length needs to be developed before a formal tag is applied. During this development stage, your students will use a variety of expressions to share the concept of length; for example, "My book is three 4-stacks long."

Do not rush the introduction of formal measurement. The measurement sticks are excellent developmental tools.

In this activity, students will come to the realization that the longer their measuring stick, the less it will take to measure the length of the object. They will also have to work out what to do if a length falls between, say, four sticks and five.

## *Using arbitrary measures.*

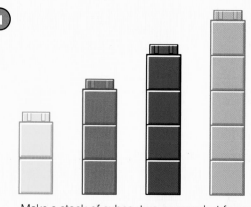

**1**

Make a stack of cubes: two or more but fewer than six. Use the same color cubes in your stack.

**2**

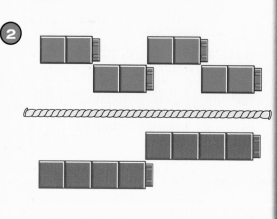

*"I have a piece of string. Look carefully. How many of your whole sticks will make the same length as this piece of string?"*

*"I notice that Mary needed four whole sticks but Enzo needed only two. How do you explain that?"*

**3** Ask the students to measure the length of various objects in the classroom; for instance, desks. Once the students have completed the measuring, review what they have found as a whole group. It should be clear to the students that the smaller the unit of measurement (i.e. the measuring stick), the more it takes to measure a length and vice versa.

## Guide to using this photocopiable resource

A variety of paper measuring sticks is provided. Use these to record the length of various items. Remember, these paper sticks are exact replicas of the real cube measurement sticks. For longer measuring sticks, glue appropriate measuring sticks together. Use the tab as the joiner.

Students may require shorter measurement sticks. In which case a longer stick may be cut to meet requirements.

The paper measuring sticks may be made to measure:

- the length of an object,
- the distance between two objects,
- the distance around an object or body part; for example, the wrist.

Do not hesitate to build ideas about height, width, length, around, along, across and between.

The concept of size includes ideas of height, width and length. In order to create an exactness, some children may cut a cube part way.

Carefully challenge this notion, explaining that to measure accurately, the sticks must be accurate and the units (cubes) within them just stay exactly the same. If a cube is cut in half we are actually creating a new unit.

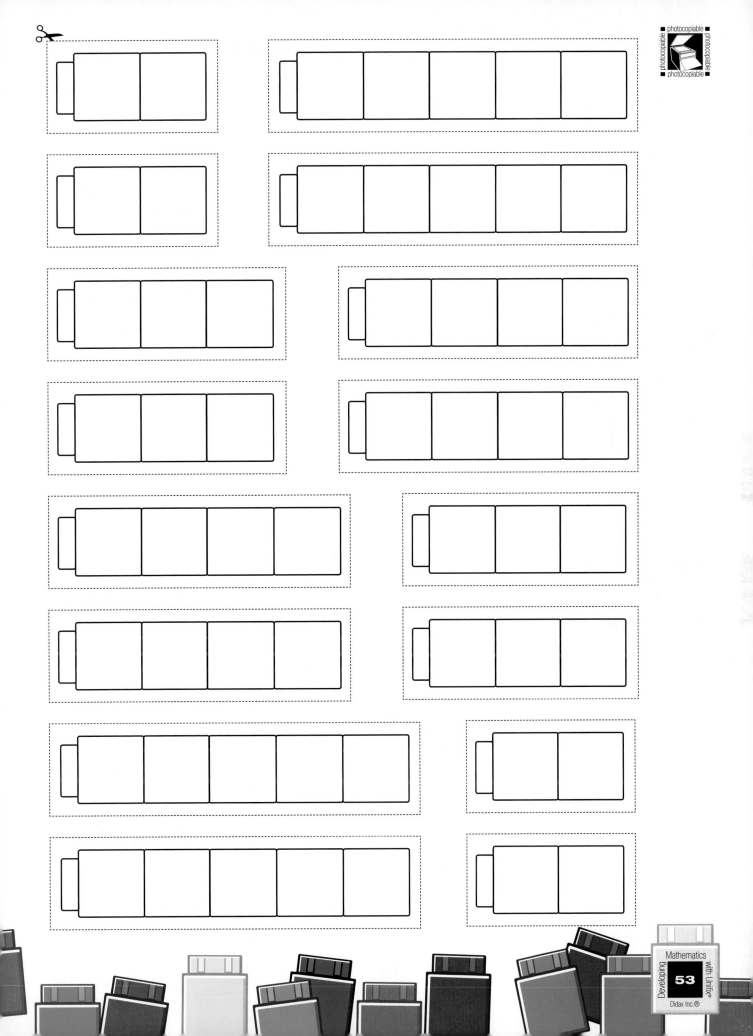

# Measurement Sticks-2

## Purpose

To realize that part of the base unit may be required in order to make an accurate measurement.

## Further development of measurement ideas

The following activities provide the opportunity to explore the process of measuring a little further. Watch how the students line up their measuring sticks.

For example, some students may measure without lining the sticks up appropriately.

Others may not keep the sticks touching when measuring.

As these issues come up, discuss them with the students.

### Measuring more accurately with measuring sticks.

**1** Set small groups to measure the lengths of various pieces of string or paper tape. Students can line up the string from the shortest to the longest (seriate) and then decide how to measure the length of the string. (Make sure some string lengths fall between the lengths of varying measuring sticks; for example, 7 cm, 9 cm, 11 cm, etc.)

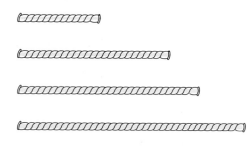

**2** In order to measure, the students will need to make some decisions about which standard measuring stick to use (a two-cube stick, five-cube stick, etc.) and then measure. As the students measure, watch how they physically place the measuring sticks along the string.

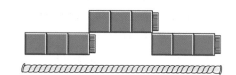

**3** Eventually the students will come across a piece of string that cannot easily be measured; for example if using a three-cube measuring stick to measure the length of a 9-cm piece of string, the students will find that one stick is too small to measure the length and that two sticks are too long.

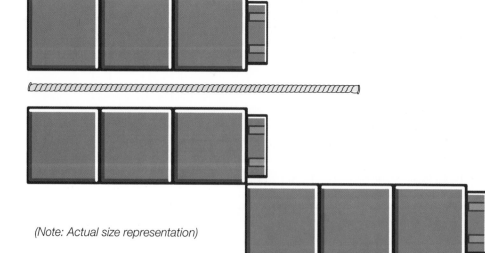

*(Note: Actual size representation)*

Encourage discussion as to how to solve this problem. When using the physical cubes, some students may suggest that you need a stick and half a stick.

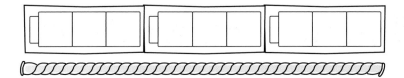

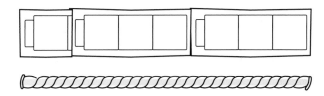

The length of each piece of string may be recorded by gluing the piece of string onto a large sheet of paper and then gluing the paper measuring sticks (that match the physical stick) next to the string. When students do this, you may note that they cut a stick into halves or parts. Use these examples to stimulate discussion when the whole group comes together.

Discuss how sometimes measurements fall between the standard lengths. (Note: This happens when using a ruler and the measurement falls between two increments; e.g. 6 cm and 7 cm.)

## *Measurement challenges*

### Find the person with the longest foot

When faced with this challenge, the students may want to use the shorter measuring sticks as it is less likely they will have problems with the foot length falling between a whole number of sticks.

This will provide you with the opportunity to discuss how smaller base units (shorter measuring sticks) are appropriate for some measuring tasks but larger base units (longer measuring sticks) are better for others. To illustrate this point, set the challenge of measuring the length or width of the classroom.

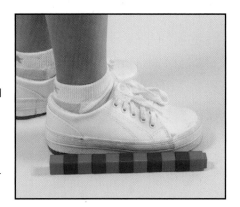

### How long is the classroom?

Allow the students to choose the length of the measuring stick they would like to use. After completing the task, ask which measuring stick the students think was better for the job. The students should appreciate that the smaller the base unit the more that are required and the more counting of the base unit that has to take place.

## Learning insights

- Students will realize they must use measuring sticks of the same length to measure an object accurately.

- It takes time and experience for students to see that the measuring sticks must be aligned end-to-end; that is, touching.

  Please do not tell them to do this; rather, challenge them to explain an example like the following:

  *Three of the same measuring sticks.*

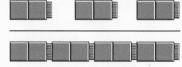

  *Four of the same measuring sticks.*

- To make an accurate measurement, an exact part of a measuring stick (in this case, one or two cubes) may be needed. A language to describe/explain this will evolve; for example, "I have used two measuring sticks and half of another one."

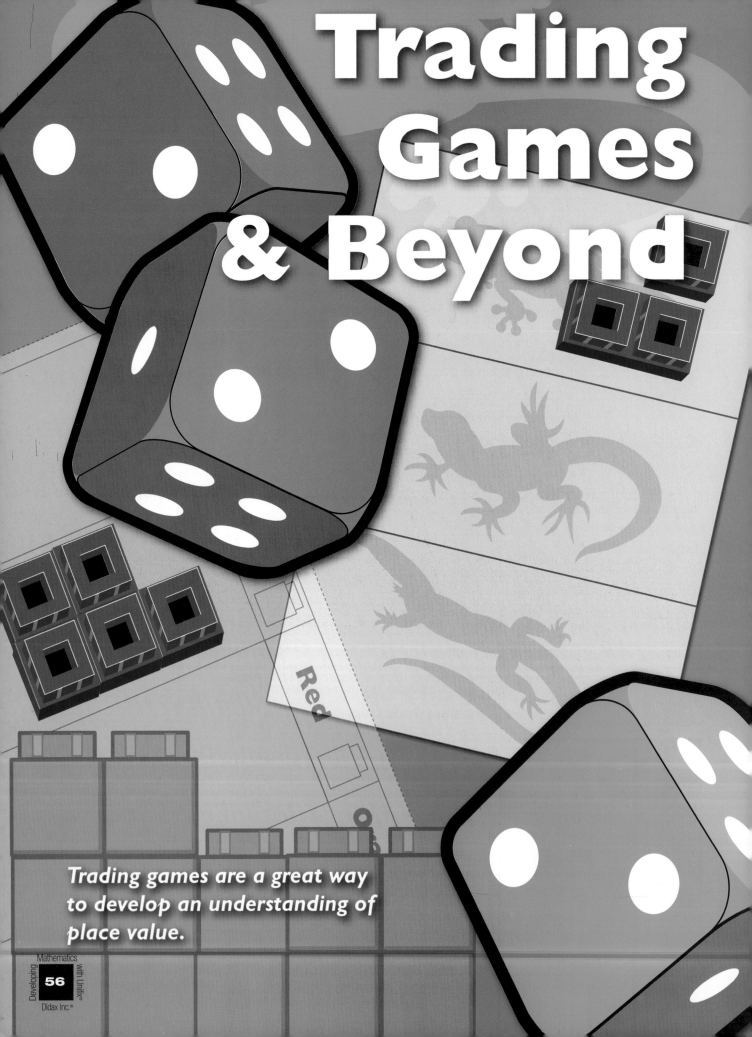

# Trading Games & Beyond

*Trading games are a great way to develop an understanding of place value.*

Developing Mathematics with Unifix® Didax Inc.®

## Introduction to trading games

Long before you introduce formal materials, like Base 10 blocks, you need to foster and develop the concepts of place value, the four operations and number facts by playing these trading games. As one excited teacher said: "The students find these games so easy and enthralling. I'm sorry I haven't used them before."

Played with a sense of fun and adventure, the students do not realize the depth of mathematical understanding in which they are involved. And, as the students begin to formalize those understandings you will be amazed at the number understanding the students have developed.

We recommend that you introduce the trading games after the students have been at school for about six to eight months. Once the students have mastered the idea of trading, leave the idea for about three weeks. During this time, the ideas will blend into the student's experiences and when you return with a "remember when ..." the students will quickly recall their trading activities.

## These organizational tips will assist you

For the first few introductory sessions with a group of students, organize whole-class exercises. Younger students may sit in a circle on the floor so you can observe how they are handling matters. Usually, older students do not need such close supervision, but we prefer to know the majority of the class has attained in-principle level of understanding of the games.

Initially, teach the students using the simple trading board. Then you can introduce them to the exciting adventure into Lizard Land!

Copy and laminate sufficient boards for every student. One teacher introduced the games on photocopied sheets. When the students knew the games, they were invited to decorate a new trading board before it was laminated. These boards became personal property which the students could take home to teach their parents how to trade.

Once familiar with the games, the students may break into groups of three or four. For a series of games, one of the students may be appointed the banker. All transactions must be passed through the banker. Personal bank books may be kept, but do not insist on formal presentation of the transactions. In time, every student will have a turn a being the banker.

Negotiate with the teacher of another class to have some of your banker-type students teach the games to that class. Bankers may wear a badge which entitles them to bank (teach) for a week.

Once the games are well-known, play Lizard Land for ten minutes before a break. Keep a progressive score.

**Purpose**

Developing trading (place value) understanding.

# Welcome to Lizard Land and its Money System

Have the students sit in a circle on the floor and act out getting on an airplane, fastening seat belts and taking off on a big adventure.

*"OK! All aboard our friendly jumbo jet … we're off on our fantastic adventure. Are our seat belts fastened? We are flying to Lizard Land."*

is called a gecko

is called a lizard

*A goanna is a type of lizard found in Australia that can be up to five feet long.*

Unifix® Cubes are used as the base of the currency. In Lizard Land, we call one cube a gecko. Geckos may be used to build lizards, goannas and even crocodiles—the latter would mean you have become very rich!

Now, because the money system in Lizard Land is based on fours, when you have earned four geckos (or four cubes), these are stacked to make a lizard.

And four lizards make a goanna! Tie the four lizards together with a rubber band; do not stack them end-to-end as they need to clearly show 4 x 4 instead of one stack of 16. Encourage discussion about this most important concept.

It is very important that these fundamental rules are well understood by the students.

Each student will have a trading board (see page 62). Later on, dice will be distributed to each group. The toss of the die will determine the number of cubes you can earn each round of play. This is a most important stage: do not skip any steps.

Create a role play. The students will happily go along for the ride.

The initial stages are important so, to assist, a script outlining what to say to the students (and some possible responses) has been provided.

| Teacher: | Let's visit a fantasy land called Lizard Land! It is in our heads, but it is real because this is our money. |
|---|---|
| | (Show some Unifix Cubes.) |
| | Ah, yes! You have to earn your money. For each clap I make, you can collect a cube. |
| | (Clap.) How many cubes do you have? |
| Students: | One. |
| Teacher: | I beg your pardon? |
| Students: | One. |
| Teacher: | I did not hear correctly. One what? |
| Students: | One cube. |
| Teacher: | Yes! One cube. One by itself does not mean anything. (Clap, clap.) |
| | How many do you have? |
| Students: | Three. |
| Teacher: | I beg your pardon? |
| Students: | Oh, yes! Three cubes. |
| Teacher: | In Lizard Land there is a very important rule. When you have four or more cubes you can make a stack of four cubes. (Clap) |
| Students: | I have four cubes. I can make a stack. |
| Teacher: | Excellent! Have all of you stacked your cubes? Show your neighbor. I'm so pleased. Now, in Lizard Land we have a special name for the stack of four. We call it a lizard. Got it? A lizard. |
| Students: | A lizard. |
| Teacher: | Make two lizards for me. Have you all got two lizards? |
| | This is excellent. |
| | (Clap, clap.) What do you have now? |
| Students: | Ten cubes. |
| Teacher: | Try again, please. |
| Students: | Oh yeah! (A toothless grin widens to split a face.) I've got two lizards and two more. |
| Teacher: | How best could you say that? |
| Students: | Oh! Two lizards, two cubes. |
| Teacher: | Do you all agree? |
| Students: | It could be two-two. |
| Teacher: | Explain that for me. (And on will go the learning conversations.) |

Please do not tell the students right or wrong. Allow them to make up their own minds. If a student has a misconception it will not take many more rounds of this game before the misunderstanding has been rectified.

Play this game until you reach three lizards and three cubes. Your main aim at present is to ensure the students know how to trade, four-for-one.

At this stage, let the game rest. About three weeks later, with the trading boards prepared, revise the Lizard Land ideas and role-play with the students that they are going on another great trip—to Lizard Land.

# Lizard Land Money Game

### Purpose

To develop the addition concept.

Carefully introduce the Lizard Land money game. You will need Unifix® Cubes, dot dice and trading boards (page 62). Follow the script to avoid missing vital stages. This game will need to be played several times to ensure proficiency before moving on.

*Welcome to Lizard Land. In Lizard Land you can live a very happy life and earn plenty of money, very easily. Today, we are going to learn how to do that. Please follow these instructions closely so you will understand*

**1**

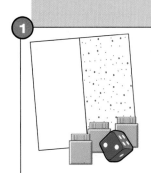

*Because we are visiting Lizard Land, this cube is called a gecko.* Identify the cubes, die and the practice trading board. Remember to keep chatting about Lizard Land as you indicate the following procedures.

**2** Rule: Geckos stay in the shade.

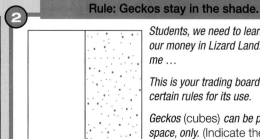

*Students, we need to learn how to use our money in Lizard Land. So, follow me …*

*This is your trading board—there are certain rules for its use.*

*Geckos (cubes) can be put in this space, only.* (Indicate the shaded right-hand space.)

**3**

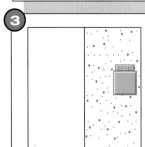

*Take one gecko and place it in the correct space.* (Check.)

**4**

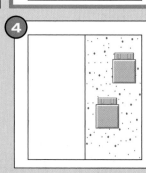

*Now, take another gecko and put it in the correct place.* (Check.)

*How many geckos altogether?*

*Two geckos!*

*Excellent!*

**5**

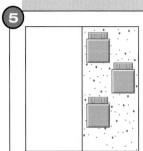

*… and a third gecko.*

Please take this stage slowly and patiently. Make sure students know where to place the geckos. You may have to carry out this procedure more than once.

**6**

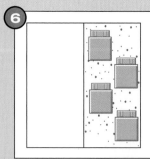

*… and another gecko.*

**7** Rule: Four geckos make a lizard.

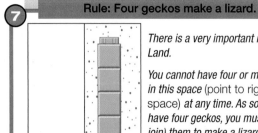

*There is a very important rule in Lizard Land.*

*You cannot have four or more geckos in this space* (point to right-hand space) *at any time. As soon as you have four geckos, you must stack (or join) them to make a lizard.*

**8** Rule: Lizards stay where there is no shade.

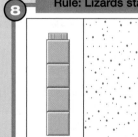

*A lizard can stay only where there is no shading on your trading board.* (Indicate the left-hand side of the board and ensure all the students have the lizard placed in the correct position.)

*Let's check: What happens if we get four geckos? Where does the lizard go?*

**9**

*Let's go on earning more money! Add another gecko. What have we now?* Encourage a (polite!) chorus.

*One lizard and one gecko.*

**10**

*Good! Add another gecko. Now what do you have?*

*One lizard and two geckos.*

**11**

*Add another gecko.*

*One lizard and three geckos.*

**12**

*Add another gecko …*

*One lizard and four..or..or Hey! Hold on! We have four geckos. But, ah! um! four geckos make a lizard. You can't have four geckos in that space.*

*Now, look how rich you are! Quick, get your trading board looking correct. Yes! We have two lizards.*

**13**

*Excellent: I think you have done very well. Let's have one more try; but, this time I will clap my hands. If I clap twice, you pick up two geckos … and so on.*
Play the game silently: the only sound being the clapping of your hands. Watch for those students who have gained mastery.

● ● ●

**14**

*By now, we have reached 3 lizards and 3 geckos. This is an impressive collection three lizards and three geckos. What will happen if I earn another gecko?*

Take all the suggestions possible and put them on show for the students to ponder. Encourage their thinking.

**15**

*If I get another gecko I will have four lizards.*

*You can't have that. You can't have four in a space and, hey, there is no space for it, anyway.*

*Any suggestions?*

## Suggestions

It could go on the edge of the board.

That's a good idea. But it looks a bit messy. How can we make it look like a group of four lizards?

Hey! What if I put a rubber band around them?

Great, but what is it called?

You can't make a big stack because that will look like a big stack and not four lizards.

I think a goanna, because it is bigger.

Oh! Wow! That's getting big. What would be the next one? Ummm…four goannas.

What about a crocodile?

Oooh! A four-goanna-crocodile: that's big.

This is a typical student-centered conversation around this stage of introducing the ideas. On one memorable occasion, this game of sizes continued: "python, iguana, alligator, boa constrictor, dinosaur. Hey, but, what sort of dinosaur?" So, they listed all until the students were satisfied that long-neck was the biggest of all. "Hey, but, we'd never have enough geckos to make a long-neck? That's huge!"

# Lizard Land Trading Board – 1

Didax Inc. ®

# Lizard Land Tax Game

## Purpose

To develop the concept of subtraction.

### From experience

Practice the game until the students demonstrate competency, but do not try to force this competency on students who do not grasp the ideas initially.

### *Moving to the E stage*

Encourage the keeping of records. Students can write down the transactions as they occur. This is characteristic of the E (explain) stage. Do not be concerned about formal representation, simply ensure the students can accurately explain what they have written. Have faith that the students know what they have recorded. Try questions like this:

You have one goanna and two geckos:

How many lizards do you have?

How many geckos altogether?

You have nine lizards and three geckos:

Arrange the collection correctly. How many goannas do you have?

This game promotes the concept of subtraction. It is at this stage we introduce the words *compose* and *decompose*. As we play the games, we will use the words to describe our actions. Do not introduce these words as separate ideas; simply mention them during the game. The students will have no problem.

This is a whole-group activity. The students are seated on the floor with their trading boards and Unifix® Cubes. You also need 2 different colored dice; one to add, one to subtract.

**1** *Hey, students! Did you know that every person living in Lizard Land has to pay tax?* (Expect groans.) *That's right, just the same as your family, they have to pay tax. Today we are going to play the tax game. First we have to earn some money. I'll throw the die.*

*Three!*

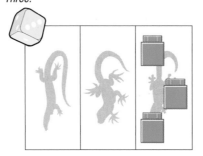

**2** Next throw.

*Four!*

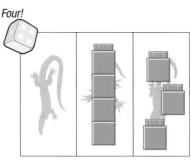

*I expect you all have one lizard and three geckos. Great! Now is the throw on which you have to pay tax. Pay back to the pile of cubes* (Later we will call it the bank.) *the number thrown on the die.* (This time throw a two.)

**3** *Two!*

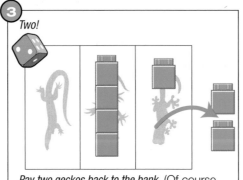

*Pay two geckos back to the bank.* (Of course, there will be fun moans.) *How much have you left on your trading board?*

*One lizard and one gecko*

**4** *Five ...*

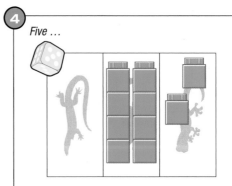

*That will be two lizards and two geckos. Agree?* (Throw the die again.)

**5**

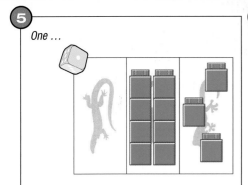

One …

*Have you all got two lizards and three geckos on your trading boards?*

*Yes!*

**6**

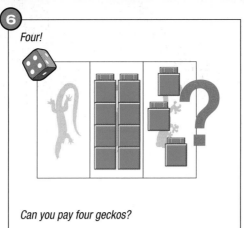

Four!

*Can you pay four geckos?*

*No, I only have three geckos.*

**7**

Continue with this type of questioning until a student says something like … *"There are four geckos in a lizard. We could use some of them."*

| | |
|---|---|
| Teacher: | *What do you think of that idea?* |
| Students: | *Pretty good.* |
| Teacher: | *Why?* |
| Students: | *Well, if I took one gecko off the Lizard, I wouldn't have a lizard any more, but I would have enough geckos to pay my tax.* |
| Teacher: | *Show me what you mean.* |
| Students: | *See. There is the tax money and this is what is left.* |

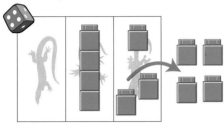

Ask another student to explain how he or she would solve the problem. Invite the students to challenge or accept the idea.

The students can then play without supervision. Make rules for winners; for example, the first to have a goanna on the trading board wins; or after four rounds, the person with the most on the trading board wins.

## Set up a shop in Lizard Land

Just like any shop, place prices on the products. But the prices will be Lizard Land prices. The students can design an appropriate symbol to indicate Lizard Land money.

## Earn the pay

As students arrive each morning, two (or three) throws of the die provide them with money which they can add to their account or spend. Additional money may be earned by doing odd jobs in the classroom.

## Teacher created situations

Without warning, tell the students they need to make a donation to help a family in need: "Please place your gecko donations in the box!"

**OR**

"We have found a treasure chest of geckos. Let's share them evenly."

## Some sound teacher talk

Ensure complete understanding. NO matter at which level these games are introduced, always carefully ensure the fundamentals are understood. The benefit of this range of games is that the students can actually see what is in each place. For example, a goanna contains four lizards and each lizard contains four geckos, making a total of sixteen cubes.

### Children must talk the moves

Encourage students to talk the moves through. As each amount on the trading board changes, the students can say something like "I have three lizards and one gecko" or "I have to pay five geckos tax. That means I have to decompose a lizard so that I have enough geckos" or "I can compose a lizard with these five geckos. So I will have one lizard and one gecko."

This talking-it-out is essential, so do not expect to play these games in silence.

### Provide practice over periods of time

There is no need to rush these games but it is essential that they are played frequently to ensure the students are competent. We prefer that they are played solidly for two or three weeks, leaving another five or six weeks before returning to them. Think of this as an intermission approach: three weeks on, six weeks off. It works!

### Concept building

Now you will have recognized how many concepts we are encouraging the students to build. By avoiding the traditional introduction to number and the operations, you are providing very appropriate experiences on which the more formal (say, abstract) aspects of mathematics can be built.

**D T E S**

## Purpose

Developing trading (place value) understanding.

Using trading games to develop the next step in understanding our number system.

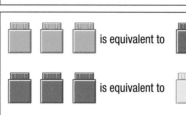

*Students, I wish to take you to another land today. This land is called Hawaii 3-0. Like Lizard Land, Unifix® Cubes are used for money, only here we start with orange cubes.*

Watch the students quickly sort out the orange cubes.

### From experience

Halfway through a student's second year at school would be a great time to introduce Hawaii 3-0.

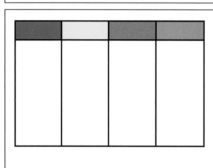

is equivalent to

is equivalent to

is equivalent to

*Yes, you earn money in Hawaii 3-0, but you are paid in orange cubes. Now, as soon as you get three orange cubes you are asked to trade them for 1 red cube, because this is a land of three; you cannot have three or more of any one color in this country. And, when you have three red cubes you need to trade them for one yellow cube and so on.*

*Wow! We will have to watch carefully.*

Distribute the Hawaii 3-0 trading board (see page 69).

*You can see the system. First, I wish to show you how people in Hawaii 3-0 trade their cubes. With their three cubes, they go to the bank and ask to be able to trade their three orange cubes for one red cube. The bank will do that.*

*Where is the bank?*

*We play imaginations. One of you will play the role of bank manager. Now this is how you do it.* Demonstrate the role-play as outlined below. After the demonstration, *Let's practice with your neighbor.*

The students practice while you supervise. *Excellent!*

## Banking role-play

We make a real fuss of the banking transaction because it is so important that the students know what is happening. Physically, they are exchanging three for one and, for many young students, that is hard to take! As you know, offer a student a larger coin of lesser value and a smaller coin of higher value and, often, it is the larger coin that is selected.

So, in role-play mode demonstrate the process with one of the students acting as customer.

Note: The role of banker should be rotated regularly.

| | |
|---|---|
| Banker: | *Ah, good morning, ma'am. What can I do for you today?* |
| Customer: | *I want to trade my money, please.* |
| Banker: | *Why?* |
| Customer: | *I can't have three of the same money in Hawaii 3-0.* |
| Banker: | *That's correct. What would you like?* |
| Customer: | *A red cube, please.* |
| Banker: | *Certainly* |

(Hand across the red cube, and receive the three orange cubes at the same time. Simultaneously! This is so important, reinforcing the ideas that three orange cubes are the same amount as one red cube.)

*There you are, ma'am. Are you satisfied?*

| | |
|---|---|
| Customer: | *Yes. I have the same value money with the red money as I had with the orange money.* |
| Banker: | *Goodbye.* |

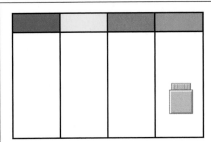

*Now, let us practice to see if we can manage our money in Hawaii 3-0. Each clap I make will earn you an orange cube. Clap. Place your cube in the correct place on the trading board. What do you have?*

*One orange cube in the orange column.*

*That's good understanding.*

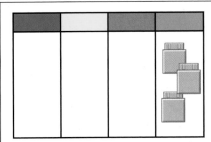

Clap, clap

*We have three orange cubes. We have to go to the bank for a trade for a red.*

*Excellent. In you go.*

The students trade their cubes.

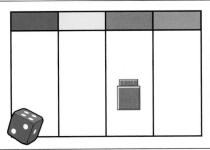

As in Lizard Land, introduce the die, reminding the students that one dot on the die means you get paid one orange cube, never any other color. As a whole group, the game continues until blue is obtained.

Occasionally, you might notice a student will stack the cubes as in Lizard Land. This is not necessary.

A few students think that once red has been reached the remainder of the game will be played in red.

About this stage a student, after throwing a 4 on the die, might request from the banker, "I want one red cube and one orange cube, please." The banker has probably said "You can't."

The teacher must step in at this point to check if the student understands what he/she is doing. If the student convinces the teacher that 4 orange cubes equals 1 red and 1 orange cube quietly say "Carry on." Never teach (tell) this stage, all students will soon realize it and a new understanding will emerge.

Students may draw cubes, make tally marks, write words or write numerals. As long as it makes sense to them, that is okay.

Remember to keep the writing and talking alive.

## Playing Hawaii 3-0 in groups of three to four students

Each group appoints a banker who controls the game. Players take turns throwing the die and the banker distributes the orange cubes. The banker will control the trading activities. The game ends when a student reaches blue. The banker is replaced by that student and another game begins.

## Keeping records

Request that the students keep a record of their transactions on paper. Do not give any more instructions than that. Allow the students to record as they see fit. Then ask the student to explain what he/she has recorded. As long as the student can explain satisfactorily, be happy. That student is well on the way to fully understanding the idea of place value.

# It's Tax Time in Hawaii 3-0

### Purpose

To consolidate the addition concept and to develop subtraction.

*Students, remember when we had to pay tax in Lizard Land? Well, the same happens in Hawaii 3-0. The government requires tax and so has ordered that on every third throw you pay back to the bank, rather than keep the money for yourself.*

This will be greeted with loud moans. Do not discourage them.

*Let's practice paying tax. Place your trading boards in front of you.*

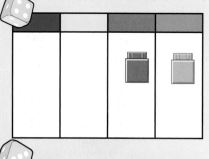

A student throws the yellow die and calls *four*. By now, most of the students will not need the banker to trade.

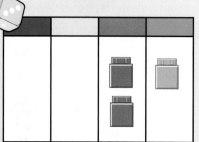

Next throw … *three*

*Aha! It is now tax time. We are going to pay the next throw in orange cubes back to the bank.*

Throw the die.

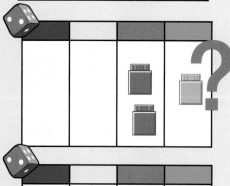

Two! Uh-oh!
What's the problem?
*We don't have enough orange cubes.*
What could you do about it? You were paid in orange cubes, remember.
*But they are red now.*
So?

I know what we could do. We could decompose the red cube into three oranges. Just like we did in Lizard Land.

*Try it to see if it works.*

The students frantically trade to discover they have two orange cubes to pay back to the bank and still have one red cube and two orange cubes remaining.

Play the game until someone earns a blue cube.

As they gain experience, the students are requested to keep records of their transactions. As long as these records make sense to the student and he/she can explain them sensibly to you, they are acceptable.

## It's bonus time in Hawaii 3-0

Students, the country of Hawaii 3-0 has just had the good fortune to find a very rich gold mine. So the government said it will share some of the money with you. You can double your collection. Double the amount you have on your trading board.

Good work. I can see you have doubled your collection very well. Now, your bank balance needs to be made legal. Please explain what you need to do.

## Now the people of Hawaii 3-0 need help

Invent a story to wrap around this idea.

Students, the government is asking for help because there has been a terrible hurricane which has caused damage to parts of the country. (divide) Make three equal groups out of your trading board money. One group will help build new houses, another group will help feed the people and the third group you may keep in your bank balance. If there is anything left over after you have divided your money into three equal groups, you may keep that amount.

Teacher Ben played the games for about three weeks each term. As each student arrived at school, he/she received a cube token. It could be one of the four cube colors. Its value could be added to the personal bank account of the student, allowing an accumulation of funds. During the day, auction items were sold and various bonuses were offered. Never was the money of Hawaii 3-0 used for rewards or punishment. The students were very eager to control their private bank accounts.

# Hawaii 3-0 Trading Board

| Blue  | Yellow | Red | Orange |
|---|---|---|---|

# Tens Land

**Purpose**

To take all the concepts developed in the trading games to Base 10.

## *Using trading games to create an awareness of Base 10.*

**1** Play the trading games as previously experienced in Hawaii 3-0 in Base 10. Make the rule *ten cubes must be traded for one cube*, which is then placed in the next column of the tray.

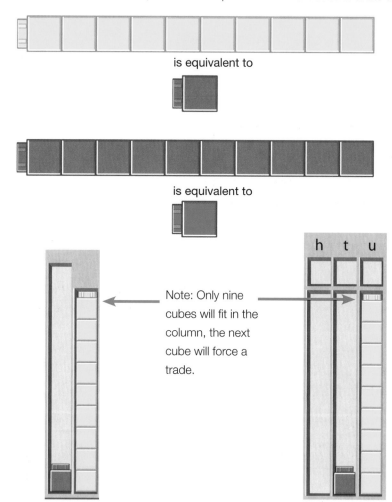

is equivalent to

is equivalent to

Note: Only nine cubes will fit in the column, the next cube will force a trade.

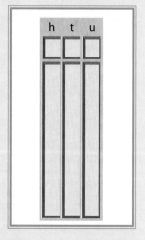

Tens and units tray

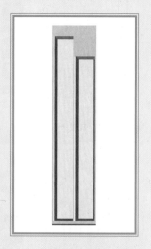

Hundreds, tens and units tray

Using these trays instead of the trading board, the games as described in Hawaii 3-0 may be played, except we are playing in Base 10, the decimal system used throughout most of the world. Note the columns of the tray will hold only nine cubes, the tenth cube forcing a trade (10 cubes become 1 cube).

### Message to teacher

Here you will see the fundamental basis of place value and the decimal system.

10 cubes (10 lots of 1) makes 1 ten in the next column

10 tens (10 lots of 10) makes 1 hundred in the next column

… and so on.

So 1,000 may be described as 10 lots of 10 lots of 10 lots of 1.

## ② Cube values

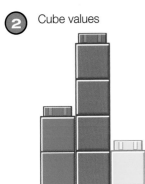

- How many tens altogether? (24 tens)
- How many tens in the tens column? (4 tens)
- How many units altogether? (241 units)
- What is the value of the cubes in the hundreds column (2 hundreds NOT 2).

H   T   U

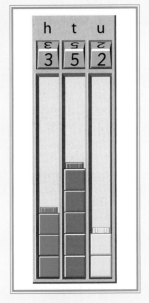

Number indicators show the symbolic representation of the cubes in the tray.

For some students, this stage will be the next logical step in their understanding. Other students may not be ready to use the number indicators, so do not pressure.

Around this stage of development, students are ready to symbolically record their ideas.

## ③ First to 100

### You will need

One hundreds, tens, units tray per student

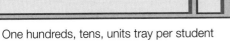

Unifix® Cubes

10-sided die

### Method

Take turns throwing the die and collect yellow cubes (units). Play the trading game until a player reaches a total of 100 or more. Each student keeps a record of each move; this written record will match the collection on his/her respective tray.

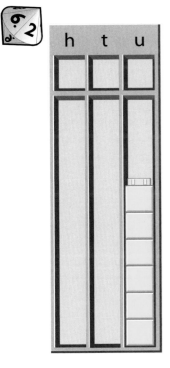

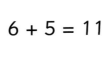

$$6 + 5 = 11$$

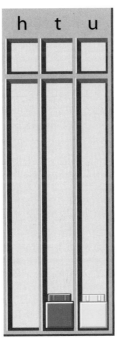

# Discovering 1 to 100 with an Operational Grid

## Purpose

Students will count beyond 10.

Students will skip count in multiples of 2, 3, 4, ...

### *Exploring a 10 x 10 matrix (1–100 grid).*

**1** **Count out 25 cubes**

Place the cubes as a group on the grid. Show at least five different ways to place the cubes as a group.

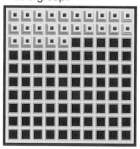
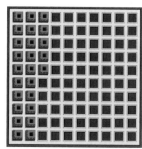
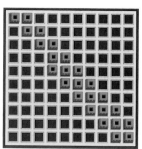

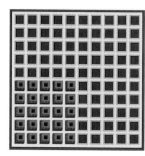
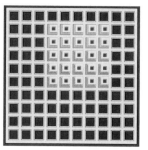

1–100 Operational Board

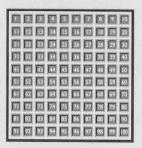

1–100 grid with point of origin at top left

We first encountered this piece of equipment when the students were exploring pattern (pages 28–35). Many of the patterns the students created when they built trains can be reproduced on the grid. Even though this is a 10 x 10 matrix, the students can create other matrices on this grid; for example, a 7-column, 6-row matrix. Experiment with these ideas.

Number indicators, number tablets and window markers are designed to fit in each cell.

The operational board may be used with or without the 1–100 grid.

**2** **Count by twos**

On every second cell, place a cube, preferably the same color.

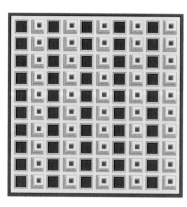

Now carefully lift the grid with the cubes still in place. Slip the 1–100 grid into the tray and replace the grid.

What do you notice? Explain.

Count by threes: Use the same procedure to discover the threes pattern.

### 3 Counting by twos from a different starting point

Still counting by twos (or threes), start at the first cell. Explain the pattern.

### 4 Counting by twos (or threes)

The 1–100 grid is placed in the tray so the numbers are evident. Using the numbers to guide them, students count by twos to 100.

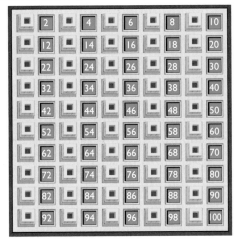

### 5 Counting by twos and threes at the same time

Use one color to mark the twos and another color to mark the threes. There will be occasions when a threes cube will need to be stacked on a twos cube.

We are sure some students will count by four and then five and so on. Allow them to discover some interesting ideas; for example, some cells will have no cubes in them.

Many teachers play the counting (factor) game of *Fizz-Buzz* with students. The activities listed may be called *Fizz-Buzz* with manipulative materials.

In *Fizz-Buzz,* a small group of students (around 10) stand in a circle. The leader chooses a multiple, take 3 for example. Each person in the group takes turns, 1, 2, 3, 4... that is one person says one, the next two, but the third person says buzz because the chosen multiple was three

Counting continues: 4, 5, buzz, 7, 8, buzz.

If a player says 6 or 9 instead of buzz, he/she is out of the game. The game may be extended to include two multiples; for example, 3 (fizz) and 4 (buzz). In this case the students on twelve would need to say "fizz buzz."

The students will be exploring lowest common multiples in an informal manner.

## Purpose

Students will relate patterns made with cubes to number patterns.

 **1** Create an AB pattern train using about 30 cubes. Place the train in the 100 track gutter.

Discuss what can be observed; for example, the relationship between the pattern and the numbers.

**2** Place an AABB train on the 100 track.

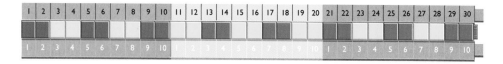

Discuss worthwhile observations:

* How many AABB groups make 12? Explain.

* From what you observe, will it be AA or BB opposite 47 and 48? Show why.

* How many AABB groups are needed to reach 20?

Repeat the same challenges using an ABC pattern.

1 to 100 track

## Describing the piece

Ten alternating grey and cream strips with a gutter designed to hold Unifix® Cubes interlock to create a 1–100 number line.

Numbers on each strip are labeled consecutively and opposite this series the numbers 1–10 are labeled. In itself, the assembling of the strips is a worthwhile experience. We recommend that the 100 track is used in conjunction with the operational grid.

The track is designed to be used as a complete 1–100 model; however, it helps to develop students' number awareness if shorter sections are used.

**3**

Create this train.

Place it on the 1–100 track with the first block alongside 1. Therefore, the last block will be opposite 15. Ask these questions and encourage explanations for the responses given by the students:

* How many groups of 3 cubes are placed on the track?

* There are 15 cubes: How many groups of 3 cubes make 15?

* Extend the train so the last cube ends at 20. Explain what you see.

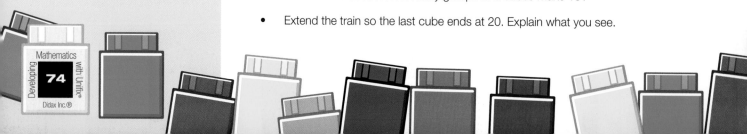

Keeping the same pattern, double the length of the train.

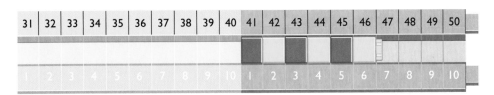

- Starting at 1, where is the end of the train?

- How many groups of three are used?

- How many groups of three are there in 30? Show why.

Use the same type of questioning for AB, ABC, AABB and AAAABBBB trains.

**4** Starting at 41, place 3 lots of AB on the 1–100 track.

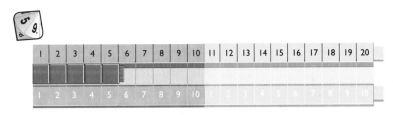

- Where is the end of the train?

- Start at 77. Where will the train end?

**5** Number line race

**You will need**

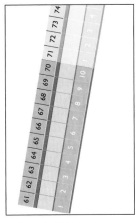

One 1–100 track per
student

Two 10-sided dice

or

Two 12-sided dice

**Method**

Each student has a complete 1–100 track and a collection of cubes. In turn, throw the dice and place that number of cubes on the track.

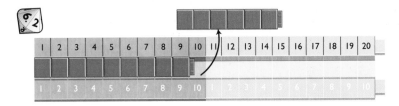

First to 100 scores a winner's point.

Introduce different colored dice which, when thrown, indicate the removal of that number of cubes. At this stage of development, expect each student to attempt to keep a written record of each move.

# The 1 to 100 Experience

### Purpose

Students will observe the relationship between the number track and the 10 x 10 grid.

***Combining ideas on the operational grid and 1–100 track to expand number awareness.***

*Compare the two alignments of the same pattern.*

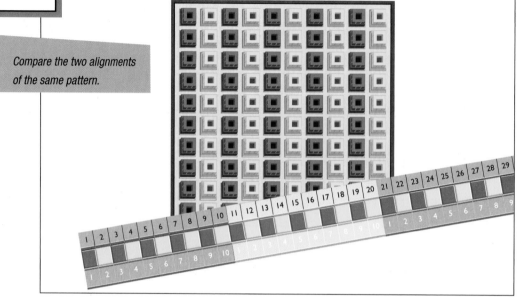

## 1 Place the cube

Focusing on two different counting pictures of 1–100—the 10 x 10 operational grid and the 1–100 number track—students locate the positions of the same numbers. The same message in a different context!

- On the grid, place a red cube on 47.
  Now place another red cube on 47 on the number track.

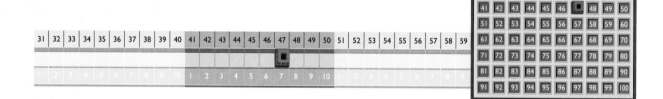

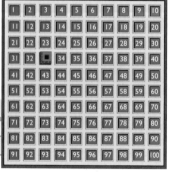

- On each, place a cube on 33.

After some practice, the students will understand the idea, so more challenging ideas will be introduced.

- Place a cube in the cell which is 2 places after 24.
- Place a cube 12 places before 75.
- Locate the position which is 23 after 59.
- When counting, place a cube nearest to 10.

Most students will place a cube on 9. Why not 11? There are several considerations to be taken into account.

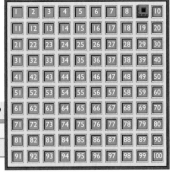

When counting, both 9 and 11 could be considered nearest to 10. On the grid, physically, 11 is a long way from 10 (a student could say 20). As teachers we need to be alert to those misunderstandings, emphasizing that the position of a number depends on its context.

- On the grid and track place a cube 10 after (or before) 28 without counting.
- Without counting, place a cube 9 before (or after) 27. How did you do that?

## 2 Count by ...

In this example we will count by three.

Using a group of three cubes each time.

Using a cube as a position marker.

Careful questioning will develop many insights.

- How many groups of 3 in 100?
- Why can't the grid or number line be filled completely?

## Extending the pattern experience

Using the template on page 41 create a variety of number grids to place in the tray.

| 100 | 99 | 98 | 97 | 96 | 95 | 94 | 93 | 92 | 91 |
|---|---|---|---|---|---|---|---|---|---|
| 90 | 89 | 88 | 87 | 86 | 85 | 84 | 83 | 82 | 81 |
| 80 | 79 | 78 | 77 | 76 | 75 | 74 | 73 | 72 | 71 |
| 70 | 69 | 68 | 67 | 66 | 65 | 64 | 63 | 62 | 61 |
| 60 | 59 | 58 | 57 | 56 | 55 | 54 | 53 | 52 | 51 |
| 50 | 49 | 48 | 47 | 46 | 45 | 44 | 43 | 42 | 41 |
| 40 | 39 | 38 | 37 | 36 | 35 | 34 | 33 | 32 | 31 |
| 30 | 29 | 28 | 27 | 26 | 25 | 24 | 23 | 22 | 21 |
| 20 | 19 | 18 | 17 | 16 | 15 | 14 | 13 | 12 | 11 |
| 10 | 9 | 8 | 7 | 6 | 5 | 4 | 3 | 2 | 1 |

| 1 | 11 | 21 | 31 | 41 | 51 | 61 | 71 | 81 | 91 |
|---|---|---|---|---|---|---|---|---|---|
| 2 | 12 | 22 | 32 | 42 | 52 | 62 | 72 | 82 | 92 |
| 3 | 13 | 23 | 33 | 43 | 53 | 63 | 73 | 83 | 93 |
| 4 | 14 | 24 | 34 | 44 | 54 | 64 | 74 | 84 | 94 |
| 5 | 15 | 25 | 35 | 45 | 55 | 65 | 75 | 85 | 95 |
| 6 | 16 | 26 | 36 | 46 | 56 | 66 | 76 | 86 | 96 |
| 7 | 17 | 27 | 37 | 47 | 57 | 67 | 77 | 87 | 97 |
| 8 | 18 | 28 | 38 | 48 | 58 | 68 | 78 | 88 | 98 |
| 9 | 19 | 29 | 39 | 49 | 59 | 69 | 79 | 89 | 99 |
| 10 | 20 | 30 | 40 | 50 | 60 | 70 | 80 | 90 | 100 |

| 0 | 1 | 2 | 3 | 4 | 5 | 6 | 7 | 8 | 9 |
|---|---|---|---|---|---|---|---|---|---|
| 10 | 11 | 12 | 13 | 14 | 15 | 16 | 17 | 18 | 19 |
| 20 | 21 | 22 | 23 | 24 | 25 | 26 | 27 | 28 | 29 |
| 30 | 31 | 32 | 33 | 34 | 35 | 36 | 37 | 38 | 39 |
| 40 | 41 | 42 | 43 | 44 | 45 | 46 | 47 | 48 | 49 |
| 50 | 51 | 52 | 53 | 54 | 55 | 56 | 57 | 58 | 59 |
| 60 | 61 | 62 | 63 | 64 | 65 | 66 | 67 | 68 | 69 |
| 70 | 70 | 72 | 73 | 74 | 75 | 76 | 77 | 78 | 79 |
| 80 | 81 | 82 | 83 | 84 | 85 | 86 | 87 | 88 | 89 |
| 90 | 91 | 92 | 93 | 94 | 95 | 96 | 97 | 98 | 99 |

Students with a strong pattern experience will see number sequences in many contexts. By widening their experience we have found the students link ideas early to the more conventional presentation of number. In fact, by encouraging experimentation and the unencumbered use of language, the traditional presentation of number to young students becomes one part of an overall vision/comprehension of how the number system works.

This understanding is translated into many aspects of school mathematics.

# Addition and Subtraction

## Expressing understanding with symbols

By now, students have developed a whole range of pattern-making skills and their counting strategies are both competent and versatile. In fact, their knowledge of the number system and the wide range of concepts involved in that understanding is strong. Developmentally, our focus has been on D (discover), T (talk) and E (explain). They are now ready to write number sentences—the S (symbol) stage of the developmental process.

Students have a strong sense of place value, realizing that the position a digit holds in any number is significant. As well, most students have developed a competent ability to manipulate numbers mentally. You will have observed this mental dexterity as the students played trading games. Constantly, we have been amazed at the many clever insights students reveal as they become involved in mathematical activities such as Hawaii 3-0.

### Converting knowledge to the written form.

We are introducing the words *compose* and *decompose* to describe the actions of addition and subtraction.

Compose: bring together, join

Decompose: break down, separate

These words described what students were doing in their heads and they readily recognized the actions. They are *teacher language* not *student language*.

**1** Compose and decompose and write the action

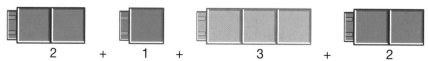

2     +     1     +     3     +     2

**Rearrange**

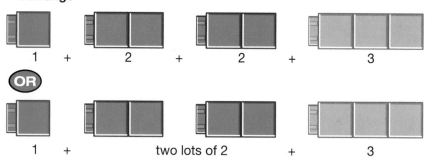

1     +     2     +     2     +     3

**OR**

1     +     two lots of 2     +     3

**Compose**

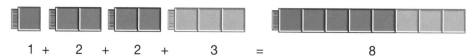

1  +   2   +   2   +   3   =   8

**Decompose**

8     –     3     =     5

**OR**

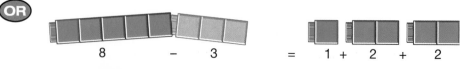

8     –     3     =   1 +   2   +   2

## 2 Compose, decompose and rearrange and write the action

### Compose

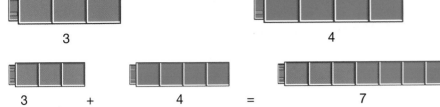

3                    4

3        +        4        =        7

### Rearrange

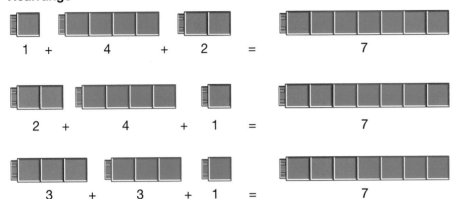

1   +        4        +   2   =        7

2   +        4        +   1   =        7

3   +        3        +   1   =        7

Discover all the rearrangements possible.

### Decompose

Remove, take, subtract 3 cubes

7        –        3        =        4

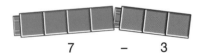

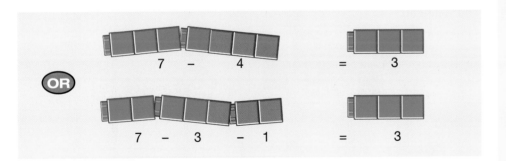

**OR**

7        –        4        =        3

7        –        3        –   1   =        3

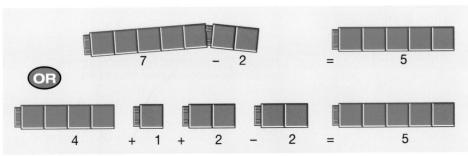

**OR**

7        –        2        =        5

4   +  1  +   2   –   2   =        5

Using this model, invent more decomposition arrangements.

## What do these symbols mean?

=

+

–

x

÷

Rather than tell the students what they mean, watch and listen to the students as they use (interpret) these symbols in their written work. Rather than correct them, identify misunderstandings and discuss them.

"… Mathematical thinking is based not on the symbols that are used, but on the meaning they are based on and come to represent …" (pg. 24)

"… the addition symbol has its origins in the latin word for 'and', et, which was used in arithmetic until it became abbreviated to + as the first mathematical symbol; literally, + is "and." (pg. 193)

Booker et al. (2004) *Teaching Primary Mathematics*, 3rd edition

This illustrates one form of subtraction. Students should be exposed to other forms.

# Composing and Decomposing

## Purpose

To relate materials (Unifix® Cubes) to the symbols.

*As formal arithmetical writing skills evolve, connection with the concrete example is essential.*

"Any new approach to mathematics must therefore recognize the fact that most children have an impressive range of mathematical abilities when they first attend school … In school, children come to grips with a new and very different kind of mathematics. I often find it useful to think of mathematics as the formal code of arithmetic, and indeed I often find it useful to think of mathematics as like a secret code, known only to those initiated into it. This code contains a number of features which distinguish it from the informal mathematics which children acquire before school. One essential feature is that it is context-free: it contains statements (such as "two plus two makes four") which are not about any situation in particular, but which can be applied to a variety of situations, old and new. It rests heavily on written symbolism (2 + 2 = 4). Both the written and context-free nature of the formal code cause considerable difficulties for young children." (pg. 168–169)

**Hughes M, 1986,** Children and Number: Difficulties in Learning Mathematics

## ① Build a stack

### You will need

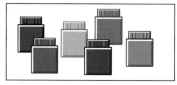

Unifix Cubes
6-sided die

### Method

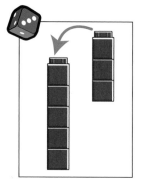

$$5+3=8$$

Start with a stack of 5 cubes of one color. On each throw of the die add that number of cubes to the stack. Record each move; for example, the first move may be 5 + 3 = 8. The first to make a stack of 22 wins a point.

## ② No more

### You will need:

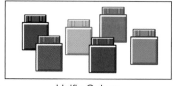

Unifix Cubes

blue die: take away

red die: add

two 6-sided dice

### Method

$$17 - 5 = 12$$

$$12 + 2 = 14$$

Start with 17 cubes. The first throw will be with the blue die, hence the recording may look like this: 17 − 5 = 12. Next throw will be added to the stack. Win the game by being the first to have no cubes.

## 3 Empty the collection

### You will need

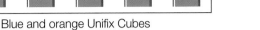
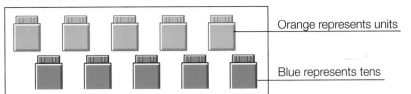

Blue and orange Unifix Cubes

Orange represents units

Blue represents tens

10-sided die

### Method

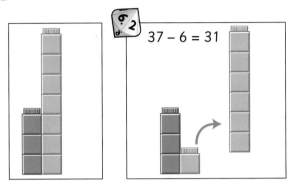
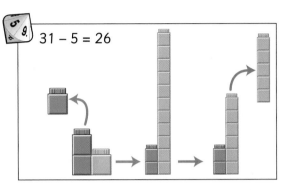

37 – 6 = 31

31 – 5 = 26

Use cubes to show 37. It is best to use a ten-sided die as this will keep the challenge moving and provide frequent trading opportunities. The first to empty the collection scores a point. Record each move thus: 37 – 6 = 31, 31 – 5 = 26. Students may wish to use a trading board (place value mat) to help keep track of the moves.

Encourage students to explain their recording.

## 4 Calculating with the hundreds, tens and units tray

### You will need

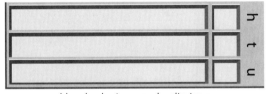
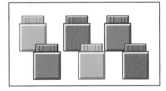
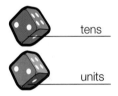

Hundreds, tens and units tray

Unifix Cubes

two 6-sided dice

tens

units

### Method

Use two different colored dice; one color represents tens, the other units.

Throw the dice and take away that amount each throw. Record your results, showing how you came to the answer.

For example, 348 – 46 = 302

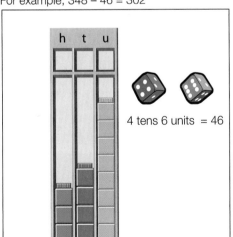

4 tens 6 units = 46

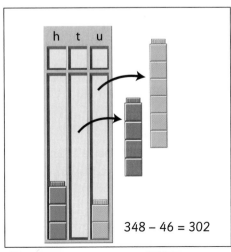

348 – 46 = 302

**Student reasoning**

- *I took away six units (ones). I knew the 4 meant 4 tens so I took them away also.*

- *I saw 48 take away 46, I knew that before so it was easy.*

- *I traded like in Hawaii 3-0. Then suddenly I realized I could have done that in my head.*

# Link and Detach

## Applying experience

It is most important that we provide students with experiences that not only foster new ideas but also strengthen those already present.

Here we present a most successful idea, one which we discovered captivated most students in the group. Please try this activity, watch what happens, then present a similar challenge later.

We used this idea as an assessment activity, not to find out what the students do not know but rather to assess the flexibility of their thinking.

*Tell all you know about 8.*

One level of thinking listed all the addition and subtraction facts:

$4 + 4 = 8, 6 + 2 = 8$

$8 - 7 = 1, 8 - 6 = 2$ and so on.

Another group introduced multiplication and division ideas, as well as mixing the operations:

$4 \times (\text{lots of}) 2 = 8$

$12 - 2 \text{ lots of } 2 = 8$ and so on.

And some students demonstrated some quite adventurous thinking and presentation.

If I share 24 marbles between 3 boys, each will have 8 marbles.

This activity provided us with a huge databank from which we were able to determine necessary and appropriate further instruction.

### A basic facts number game for two or more players.

### Setting up

Each player has a collection of 12 cubes, preferably of the same color. Link some to make a collection similar to this one.

There are two piles of cards (see page 83).

L and D cards

 L means link two pieces together.

 D means "detach" to make one piece into two pieces.

Number cards

The above collection would read

$3 + 1 + 4 + 2 + 2$

OR *(if the links were rearranged)*

$1 + 2 + 2 + 3 + 4$, etc.

### Play

Each player arranges his/her collection individually.

Initially, players choose either an appropriate number card and an L or D card. (Later, the element of chance may be introduced.)

Match the pieces according to the instructions on the cards. For example:

Original arrangement:

Select a number card:

$3 + 3 + 6$

Select an L or D card:

**L**

Final arrangement:

link

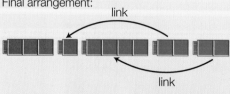

link

Each player keeps the number card and play continues until all the cards are used. Follow up by making a display of the pathways created during the game.

### Guide to using this photocopiable resource

Copy onto cardstock. Cut out each card: the number cards make one collection and the L and D cards another.

The instructions provided to play this game are just one way of investigating this join and separate concept. As you and your students experience this challenge, you will create new conditions under which to conduct it.

| | | | |
|---|---|---|---|
| **L** | 1 + 11 | 1 + 1 + 10 | 2 + 2 + 8 |
| **D** | 2 + 10 | 1 + 2 + 9 | 2 + 3 + 7 |
| **L** | 3 + 9 | 1 + 3 + 8 | 2 + 4 + 6 |
| **D** | 4 + 8 | 1 + 4 + 7 | 2 + 5 + 5 |
| **L** | 5 + 7 | 1 + 5 + 6 | 3 + 3 + 6 |
| **D** | 6 + 6 | 4 + 4 + 4 | 3 + 4 + 5 |
| **L** | 1 + 1 + 1 + 9 | 1 + 1 + 5 + 5 | 2 + 2 + 2 + 6 |
| **D** | 1 + 1 + 2 + 8 | 1 + 2 + 2 + 7 | 2 + 2 + 3 + 5 |
| **L** | 1 + 1 + 3 + 7 | 1 + 2 + 3 + 6 | 2 + 2 + 4 + 4 |
| **D** | 1 + 1 + 4 + 6 | 1 + 2 + 4 + 5 | 3 + 3 + 3 + 3 |

# The Name Game

## Purpose

To develop an understanding of "average."

## Background

The term average may apply to several different measures. These are mode, median and mean.

- Mode is the most commonly occurring number in a set of numbers.

  Sometimes you can have more than one mode.

- Median is the middle number when a set of numbers is arranged in order from smallest to largest or vice-versa.

- Mean is the value when all the numbers in a set are added and the total is divided by how many numbers were in the set.

### *A practical approach to averages.*

**1** Arrange the students into groups of seven or nine (an odd number will make it simpler for later activities). Place a pile of Unifix® Cubes in the middle of the groups and ask the students to pick up and stack the same number of cubes (all the same color) as the letters in their first name. For example, Adrian would pick up six green Unifix Cubes and join them.

**2** Ask the students to place the stack of cubes representing the number of letters in their first name in front of them. Discuss who has the longest/shortest name and which names are the same length.

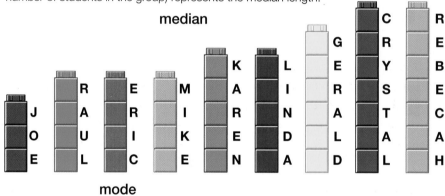

**3** Next, order the stacks from shortest to tallest. The stack in the middle (if there is an odd number of students in the group) represents the median length.

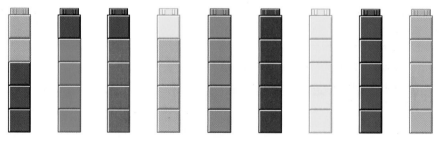

**4** The mode refers to the most commonly occurring name length. For the group shown above, the most common name length is four letters.

**5** Discuss the third type of average—the mean. Ask the students to suggest a way to make all the stacks the same height. By removing some cubes from the tall stacks and adding them to the shorter stacks you should end up with all the stacks being the same height. (At this stage don't worry if you have a cube or two left over.) The final stack height represents the mean letter length for names in the group.

**mean**

Note: If there were eight children in the group, stacks four and five would be added together and halved to calculate the median.

# Complete the Wall

## *Add to existing columns to complete the wall.*

### You will need

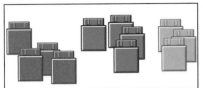

A collection of Unifix Cubes     A 6-sided die

### Background

This game will expose the students to the idea of complimentary addition; that is "Do I need to add to a number to make another?" $(3 + ? = 10)$

### Aim

A wall height and width are chosen at the start of the game. For example, a height of ten cubes and a width of eight cubes might be chosen.

Students play in pairs to build the wall.

### Method

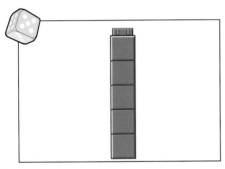

**1** Each player chooses a different color Unifix Cube to use. The first player rolls the die, picks up that many cubes and starts building the wall.

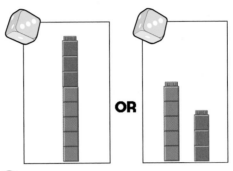

OR

**2** The second player does the same. He/She may add to the first player's column or start a new column, remembering that the wall can only be ten cubes high and eight cubes wide.

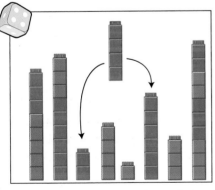

**3** As the game progresses, the students will start to use strategy when building the wall. For example, if a four is rolled a student might choose to complete a column that was six cubes high, or add to a column that was only two high.

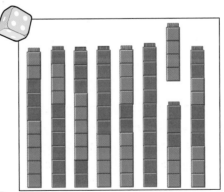

**4** The winner is the player who completes the wall.

Note: You must place all the cubes from a throw together; for example, if you throw a five you cannot put two cubes on one column and three on another.

### Variations

The height of the wall may be changed. Instead of starting with a column ten cubes high you could start with one six cubes high.

The type of die used could be altered; that is, use a four- or ten-sided die.

# Building a Graph

## Purpose

To create a simple graph.

## Background

The ability to create and interpret graphs is at the foundation of statistical literacy. Young children should collect primary (first-hand) data and then choose an appropriate way to depict the data.

Once their graphs are complete, students will be encouraged to talk and write about them.

They say a picture tells a thousand words and every graph tells a story, so the students need to be able to interpret what they have drawn.

### *Collecting data to produce a graph.*

**1** Place a large tub (1,000–2,000 cubes) of Unifix® Cubes in front of a student and ask him/her to scoop up as many cubes as possible. Young students will need to use two hands. Repeat for a small group or the whole class. While you are doing this, most students will naturally sort the cubes, generally by color, and join them.

The stacks may then be lined up.

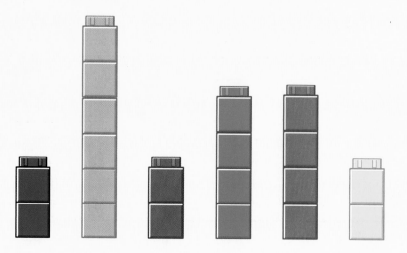

**2** Encourage the students to talk about their graphs. They will use expressions such as tallest, shortest, same, more and less when describing the graph.

**3** Pose the question, "How will we know what our stacks were like when we have packed the cubes away?"

Students may suggest taking a photo, making a drawing or writing down how many of each color cube they had. There are two drawings or graphs they might create from this data: pictograph and a bar graph.

Make sure the students write about their graphs. This will provide you with an excellent opportunity to evaluate their thinking.

## Guide to using this photocopiable resource

Use the resource to make a cube pictograph:

- A representation of the cubes is used to make a pictograph. One cube should be colored to represent each cube of a particular color. For example, if three green cubes were in a stack then three green cubes would be colored.

Use squared graph/grid paper to make a cube bar or column graph:

- Draw and label the axes of the graph and color squares on graph/grid paper to match the corresponding number of cubes.

# Number of Cubes

**Color**

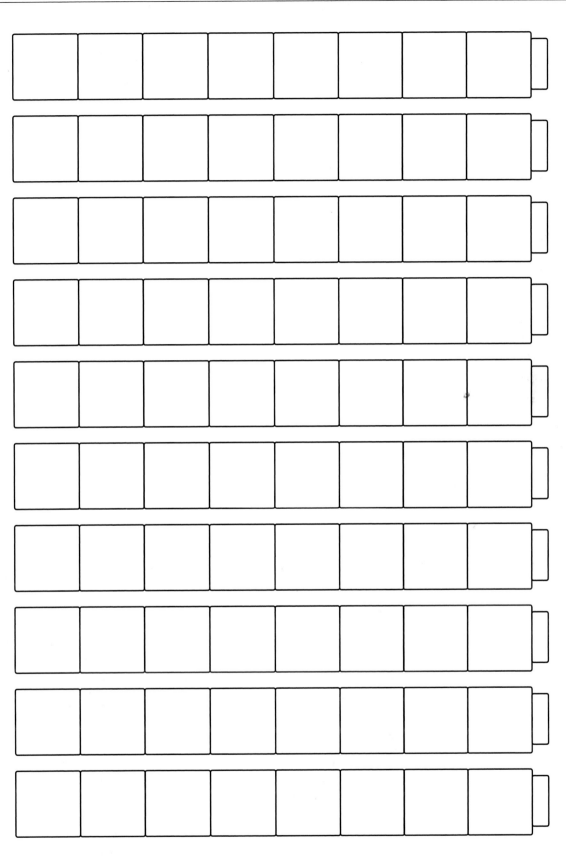

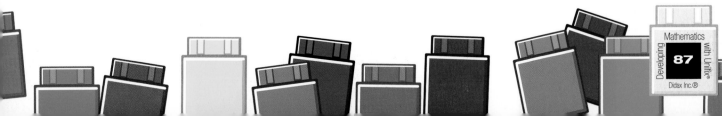

**Base 10** – the name of the number system we use. The digits 0, 1, 2, 3, 4, 5, 6, 7, 8, 9 are combined via the use of place value to form any number. Other bases such as Base 2, which uses the digits 0 and 1, are used to program computers.

**Cardinal number** – the cardinality of a set of objects or things according to particular attributes such as size, shape and color.

**Classification** – the process of arranging objects or things according to particular attributes such as size, shape and color.

**Digit** – 0, 1, 2, 3, 4, 5, 6, 7, 8, 9, are examples of digits. A number such as 297 may be described as a three-digit number.

**Number** – is the abstract idea of quantity; i.e. how many.

**Numeral** – the written or spoken symbol that represents a number; e.g. 6 or the Roman numeral VI.

**Ordinal number** – tells what order or position an object occupies in a sequence; e.g. 1st, 2nd, 3rd.

**Pattern/Design** – pattern is at the heart of mathematics. When students notice some regularity or some element that is repeated, they are noticing a pattern. Patterns come in different forms, such as number patterns, 2, 4, 6, 8 … or the patterns inherent in mats and fabric where one element is repeated in a particular way; for example, turning the element 180°. The pattern may thus be described so it may be reproduced.

**Place value** – is the concept that underpins our number system. It allows the 10 digits 0, 1, 2, 3, 4, 5, 6, 7, 8, 9 to be used to produce numbers. The position of one digit relative to another determines its value. For example, in 768 the 6 represents 6 tens or 60. Every time a digit is moved one place to the left it has the effect of being multiplied by 10.

**Seriation** – when you have two objects you compare; for example, one object might be heavier than another, which in turn means the other object is lighter. Once you have more than two objects you seriate; that is, order the objects; for example, from lightest to heaviest. Note that language changes as one object may be heavier than another but lighter than a different object in the group.

**Subitizing** – the ability to glance at a small arrangement of objects and instantly state how many are in the set.

Over the years, we have been inspired to use these materials by educationalists such as Mary Baratta Lorton, Pamela Leibeck and Kathy Richardson.

| Strand | Expectations | Correlated Activity (page) |
|---|---|---|
| **Number and Operations** | | |
| Understand numbers, ways of representing numbers, relationships among numbers and number systems | Count with understanding and recognize "how many" in sets of objects | Three Cubes on Your Fingers (p. 38) |
| | Use multiple models to develop understanding of place value/base ten | Lizard Land Money Game (p. 52) Tens Land (p. 62) |
| | Develop understanding of the relative position and magnitude of whole numbers and of ordinal and cardinal numbers | Blocks in Socks (p. 8) Introducing Value Boats (p. 41) |
| | Develop a sense of whole numbers and represent and use them in flexible ways, including relating, composing and decomposing numbers | Lizard Land Tax Game (p. 56) Composing and Decomposing (p. 72) Discovering 1-100 (p. 64) |
| | Connect number words and numerals to the quantities they represent, using various physical models/representations | Introducing Number Indicators (p. 40) Link and Detach (p. 74) Number Patterns (p. 34) |
| Understand meanings of operations and how they relate to one another | Understand various meanings of addition and subtraction of whole numbers and the relationship between the two operations | Tax Time in Hawaii 3-0 (p. 60) Addition and Subtraction (p. 70) |
| | Understand the effects of adding and subtracting whole numbers | Addition and Subtraction (p. 70) Link and Detach (p. 74) |
| | Understand situations that entail multiplication and division, such as equal groupings of objects and sharing equally | Measurement Sticks-2 (p. 46) Snakes (p. 30) |
| Compute fluently and make reasonable estimates | Develop and use strategies for whole-number computations, with a focus on addition and subtraction | Complete the Wall (p. 77) |
| | Use a variety of methods to compute | The 1-100 Experience (p. 68) |
| **Algebra** | | |
| Understand patterns, relations, and functions | Sort, classify and order objects by size, number and other properties | Making Long Trains (p. 14) Cube Stacks (p. 28) |
| | Recognize, describe and extend patterns | Snap-Clap (p. 12) Growing Patterns (p. 18) Cars in the Train Barn (p. 20) Snakes (p. 30) |
| **Measurement** | | |
| Understand measurable attributes of objects and process of measurement | Understand how to measure using nonstandard and standard units | Measurement Sticks (p. 44) |
| Apply appropriate techniques, tools and formulas | Use repetition of a single unit to measure something larger than the unit | Measurement Sticks (p. 44) |
| | Measure with multiple copies of units of the same size | Measurement Sticks-2 (p. 46) |
| **Data Analysis** | | |
| Formulate questions that can be addressed with data and collect, organize and display | Represent data using concrete objects, pictures and graphs | Discovering 1 to 100 (p. 66) The Name Game (p. 76) |
| | Sort and classify objects according to their attributes and organize data | Building a Graph (p. 78) |

| Unifix Cubes | 1–10 Value Boats |
| Ten and Units Tray | Hundreds, Tens and Units Tray |
| 100 Track | Number Indicators (50) |
| Operation Grid and Container | Operational Board |
| Pattern Building Underlay Cards (12) | 1–100 Operational Board Inset |

# Notes

# Notes

# Notes

# Notes

# Notes

# Notes